AF767174

Pattern Formation and Stability in Magnetic Colloids

Edited by Nicolás O. Rojas

Published in London, United Kingdom

IntechOpen

Supporting open minds since 2005

Pattern Formation and Stability in Magnetic Colloids
http://dx.doi.org/10.5772/intechopen.76492
Edited by Nicolás O. Rojas

Contributors
Sebastian Altmeyer, Martin Hernandez Contreras, Ricardo Peredo Ortiz, Raquel Hernandez Gomez, Chaabane Raoudha, Abiodun Opanuga, Olasunmbo Olaoluwa Agboola, Hilary Okagbue, Shiela Bishop, Nicolás O. Rojas

© The Editor(s) and the Author(s) 2020
The rights of the editor(s) and the author(s) have been asserted in accordance with the Copyright, Designs and Patents Act 1988. All rights to the book as a whole are reserved by INTECHOPEN LIMITED. The book as a whole (compilation) cannot be reproduced, distributed or used for commercial or non-commercial purposes without INTECHOPEN LIMITED's written permission. Enquiries concerning the use of the book should be directed to INTECHOPEN LIMITED rights and permissions department (permissions@intechopen.com).
Violations are liable to prosecution under the governing Copyright Law.

Individual chapters of this publication are distributed under the terms of the Creative Commons Attribution 3.0 Unported License which permits commercial use, distribution and reproduction of the individual chapters, provided the original author(s) and source publication are appropriately acknowledged. If so indicated, certain images may not be included under the Creative Commons license. In such cases users will need to obtain permission from the license holder to reproduce the material. More details and guidelines concerning content reuse and adaptation can be found at http://www.intechopen.com/copyright-policy.html.

Notice
Statements and opinions expressed in the chapters are these of the individual contributors and not necessarily those of the editors or publisher. No responsibility is accepted for the accuracy of information contained in the published chapters. The publisher assumes no responsibility for any damage or injury to persons or property arising out of the use of any materials, instructions, methods or ideas contained in the book.

First published in London, United Kingdom, 2020 by IntechOpen
IntechOpen is the global imprint of INTECHOPEN LIMITED, registered in England and Wales, registration number: 11086078, 7th floor, 10 Lower Thames Street, London, EC3R 6AF, United Kingdom
Printed in Croatia

British Library Cataloguing-in-Publication Data
A catalogue record for this book is available from the British Library

Additional hard and PDF copies can be obtained from orders@intechopen.com

Pattern Formation and Stability in Magnetic Colloids
Edited by Nicolás O. Rojas
p. cm.
Print ISBN 978-1-78985-379-7
Online ISBN 978-1-78985-380-3
eBook (PDF) ISBN 978-1-83880-275-2

For EU product safety concerns:
IN TECH d.o.o., Prolaz Marije Krucifikse Kozulić 3, 51000 Rijeka, Croatia, info@intechopen.com or visit our website at intechopen.com.

We are IntechOpen,
the world's leading publisher of
Open Access books
Built by scientists, for scientists

4,500+
Open access books available

118,000+
International authors and editors

130M+
Downloads

151
Countries delivered to

Our authors are among the
Top 1%
most cited scientists

12.2%
Contributors from top 500 universities

WEB OF SCIENCE™

Selection of our books indexed in the Book Citation Index
in Web of Science™ Core Collection (BKCI)

Interested in publishing with us?
Contact book.department@intechopen.com

Numbers displayed above are based on latest data collected.
For more information visit www.intechopen.com

Meet the editor

Nicolás O. Rojas obtained his PhD in Physics at the Université de Nice in France. His main research disciplines include nonlinear physics, statistical mechanics, biological physics, fluid dynamics, statistical hydrodynamics, thermodynamics of mixing, lubrication theory, fluid-structure interaction, theory of elasticity, superparamagnetic colloids, finite volumes method and the finite elements method. He has been a postdoctoral fellow in the Service de Physique de l'État Condensé (SPEC) at CEA Saclay, at the Group for Research and Applications in Statistical Physics (GRASP), Université de Liège, and in the Manchester Centre for Nonlinear Dynamics (MCND) at the University of Manchester. His publications are indexed ISI and in international databases. He has been referee for the Experimental Thermal and Fluid Science, Physics of Fluids and Journal of Fluid Mechanics.

Contents

Preface

This book offers a series of chapters on the topics of magnetic suspensions, statistical mechanics, thermodynamics and computational physics. Authors from countries of four continents: Africa, Europe, North America, and South America published research works derived from their studies. The readers will be confronted with new questions, solutions and answers relative to emerging topics in ferrofluids. This book can be of interest to a large audience: undergraduate and graduate physics students, engineers, scientists , and faculty members wanting to participate in magnetic liquids research. The chapters can be used as complementary material of introductory courses on magnetic mixtures. The selection of articles is suitable for the instruction of engineers and researchers in disciplines relative to magnetic colloids and for readers with technical background. This compilation can be also useful to industrial personnel and technical staff with knowledge in the area of magnetic suspensions. The presented subjects were written to offer a straightforward overview, each chapter goes into deep theoretical rationale, numerical methods and comparison against experimental measurements. The open access feature gives worldwide visibility to the book, scientists can freely acquire the chapters and interact with the available contents.

This book contains six chapters on pattern formation and stability of magnetic colloids. The topics in this book range from investigations on Taylor-Couette flows to research works on couple stress fluid and entropy generation. Analytical methods and numerical algorithms such as the Lattice Boltzmann Method (LBM), spectral decompositions, and differential transformation (DTM) methods are employed. The introductory chapter presents the cohesion energy of superparamagnetic aggregates in a colloidal suspension. The computation is performed up to 4th order employing the distance levels method. The second chapter contains an investigation on the Taylor-Couette ferrofluidic flow. The axial, transverse and oblique applied fields yielding wavy Taylor vortex flows (wTVF) and wavy spirals (wSPI) are described. In the third chapter, the authors report theoretical and numerical findings of structures and dynamical properties of magnetic colloids in thermodynamic equilibrium. A survey of the different models is provided. The fourth and sixth chapters present magneto-convection transfer in linearly and partially open cavities respectively. In both chapters LBM is employed. The fifth chapter is a research work on the convection flow of an MHD couple stress fluid in a vertical microchannel, with findings on the fluid velocity, temperature, entropy generation and Bejan number in the channel. The book offers important information to further develop novel projects in the field of magnetic colloids.

The editor acknowledges the authors of the manuscripts for their scientific contributions. Each chapter was submitted to a rigorous selection process leaving out a great amount of publication proposals. It is a pleasure to thank the editorial members that have participated in the book corrections, review and publication. The editorial team provided excellent standards that ensured the quality and relevance of the accepted articles.

Nicolás O. Rojas
Centro de Física No Lineal de Santiago,
Chile

Chapter 1

Introductory Chapter: An Example in Superparamagnetic Colloids

Nicolás O. Rojas

1. Magnetic beads aggregation

Magnetic colloids is an increasingly growth field that includes all the physical phenomena related to magnetic interactions in a colloidal suspension. Research on magnetic suspensions is extensive and of high importance in the physics community. The involved areas range from theoretical statistical mechanics and experimental physics to numerical methods such as Monte Carlo, Langevin dynamics and the Lattice-Boltzmann method. A good review of the main components of the theory of ferromagnets can be found in [1], which is used as a standard textbook in any solid-state physics course. The aim of this introduction is to straightforward lead to a concrete example in ferrofluids. I will provide a brief description as a demonstrative instance of how the research topics in this field become rapidly complex and rich in contents (more than three areas of physics combined). In the present case, the involved research areas are statistical mechanics, electromagnetism, thermodynamics of mixing and solid-state physics.

In thermodynamic equilibrium, superparamagnetic monomers at the micron scale gather forming chains [2] parallel to the magnetic field direction. In turn, chains aggregate building one-layered structures (ribbons or bundles) due to the screening effect inherited by the magnetic potential of a dipole [3] and the effective potential between chains or between a chain and a bundle. The latter was experimentally found in a quasi-two-dimensional vessel under a uniform magnetic field. A detailed study of the interaction between a magnetic particle and a chain and between two chains is given in [4]. In the latter reference, the formation process is addressed including compact packing and noncompact cases (referred as barely touching chains).

2. Cohesion energy of the aggregates

The study of formed (after aggregation) compact regular and quasi-rectangular bundles of number of particles n, width d and length s (these distances are measured in number of grains) has been performed in [5]. In addition to these cases here, a description of non-rectangular (irregular) bundles is included. Rectangular aggregates contain a number of particles $n = ds$, quasi-regular aggregates hold $n = ds + n\%d$ and non-rectangular bundles have irregular forms due to the aggregation process in thermodynamic equilibrium (see **Figure 1**). $n\%d$ is equal to the rest of the integer division n/d. Regular and quasi-rectangular aggregates of the same width cannot have the same amount of particles since the rest $n\%d$ is zero only in regular cases. Nevertheless, rectangular and quasi-regular bundles of different width can contain the same number of grains. Irregular ribbons can coincide in n with a rectangular or a quasi-regular bundle (see **Figure 1b,c**). The energies of the

IntechOpen

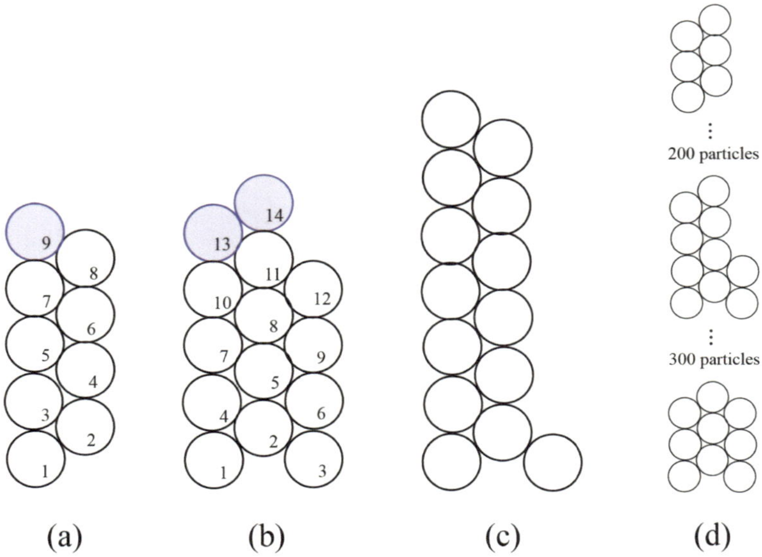

Figure 1.
Indexed compact (quasi-)rectangular bundles in the cases (a) (d, s) = (2, 4) and (b) (d, s) = (3, 4) with a total number of particles n = ds + n%d and rest particles (blue) filled. (c) Not indexed compact irregular aggregate at n = 14. (d) Non-rectangular aggregate of size n = 500 with upper width d_1 = 2 and lower width d_2 = 3.

latter cases are different. The amount of magnetic interactions in a ribbon is separated in the number of negative interactions $n_m^{(i)}$ and the number of positive bonds $n_p^{(i)}$ (relative to attractive and repulsive forces, respectively) at each order i, which is employed to compute the energy of a chain or regular bundle up to order i_{max}:

$$U^{(i_{max})}(d,n) = \sum_{i=1}^{i_{max}} n_m^{(i)} e_m^{(i)} + n_p^{(i)} e_p^{(i)} \tag{1}$$

Here $e_m^{(i)}$ and $e_p^{(i)}$ are the magnitude of negative and positive bonds at each order, respectively. The order of positive and negative interactions is relative to the distance between two grains in a ribbon [1]. Eq. (1) (see **Figure 2**) yields the explicit dependence of the aggregate energy on d and s; at higher orders the results converge to the exact curves obtained in [5] with a different method (indexing grains). Thus, the exact curves serve as a reference of the rectangular or quasi-regular ribbon magnetic energy dependency on the number of particles, although this dependence is different if irregular cases are considered. Eq. (1) is also useful to fit the exact results and to compute the energy of irregular bundles.

Quasi-regular bundles can be approximated to rectangular cases; this is called coarse-grained approximation since it does not take into account details (local groups of points) [5]. Irregular ribbons can have the same energy and number of grains of the reference (exact) curves (d = 1, 2, 3, 4), although their energy is in general located around these curves. Some cases are similar to rectangular or quasi-regular aggregates as shown in **Figure 1c**, where a non-rectangular ribbon is similar to a bundle of width d = 2; in this case its energy is equal to the energy of the regular aggregate plus all the interactions relative to the extra particle. Irregular aggregates occur also at higher n (see **Figure 1d**); the energy of a bundle at n = 500 is equal to the rectangular case n = 600 minus all the interactions relative to the 100 particles lacking at the third column. An intersection point in **Figure 2** occurs when two bundles of different width have the same number of particles and energy; the reference curves indicate that wider aggregates are more stable at increasing the number of particles, which is a plausible explanation for the experimental observation of

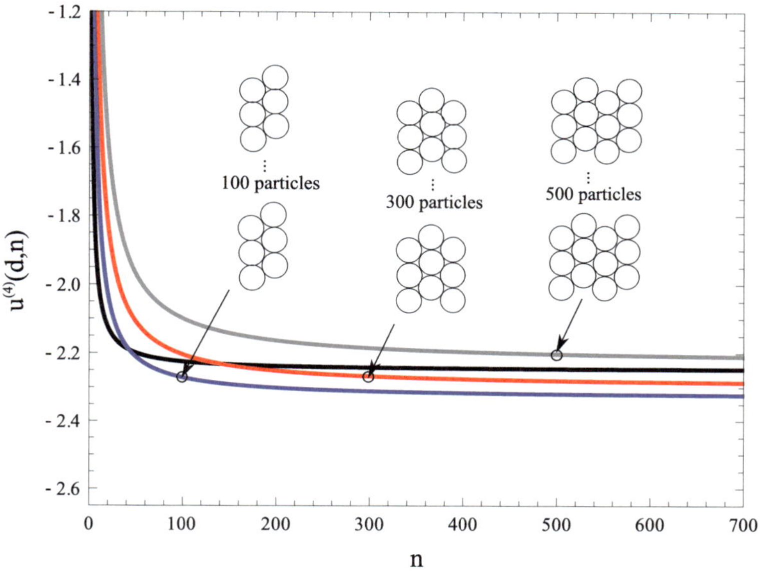

Figure 2.
Normalised magnetic energy of regular bundles at the fourth order, d = 1 (black), d = 2 (blue), d = 3 (red) and d = 4 (grey) curves. First intersection at about n = 40. Schemes of regular aggregates at (d, n) = (2, 100), (3, 300) and (4, 500) marked in circles. Convergence to the exact curves and intersections at n = 30, 113, 263 are obtained at higher orders.

ribbons in the suspension. The exact curves can be employed in statistical tools such as the standard part of the chemical potential and the equilibrium constant in order to obtain the mean length of the aggregates, which present multistability at increasing the magnitude of the imposed magnetic field.

Author details

Nicolás O. Rojas
Centro de Física No Lineal y Sistemas Complejos de Santiago, Chile

*Address all correspondence to: nicolas.rojas@email.com

IntechOpen

© 2019 The Author(s). Licensee IntechOpen. This chapter is distributed under the terms of the Creative Commons Attribution License (http://creativecommons.org/licenses/by/3.0), which permits unrestricted use, distribution, and reproduction in any medium, provided the original work is properly cited.

References

[1] Kittel C. Introduction to Solid State Physics. 7th ed. New York: John Wiley; 1996

[2] Faraudo J, Andreu JS, Camacho J. Understanding diluted dispersions of superparamagnetic particles under strong magnetic fields: A review of concepts, theory and simulations. Soft Matter. 2013;**9**:6654-6664

[3] Darras A, Fiscina J, Pakpour M, Vandewalle N, Lumay G. Ribbons of superparamagnetic colloids in magnetic field. European Physical Journal E: Soft Matter and Biological Physics. 2016;**39**:47

[4] Messina R, Spiteri L. On the interaction of dipolar filaments. European Physical Journal E: Soft Matter and Biological Physics. 2016;**39**:81

[5] Rojas N, Darras A, Lumay G. Self-assembly processes of superparamagnetic colloids in a quasi-two-dimensional system. Physical Review E. 2017;**96**:012608

Chapter 2

Interaction of Magnetic Fields on Ferrofluidic Taylor-Couette Flow

Sebastian Altmeyer

Abstract

When studying ferrofluidic flows, as one example of magnetic flow dynamics, in terms of instability, bifurcation, and properties, one quickly finds out the additional challenges magnetic fluids introduce compared to the investigation of "*classical*", "*ordinary*" shear flows without any kind of particles. Approximation of ferrofluids as fluids including point-size particles or, more realistic fine size particles, the relaxation times of the magnetic particle, their interaction between each other, i.e., the agglomeration and chain forming effects, and the interaction/response between any external applied field and the internal magnetization are just few examples of challenges to overcome. Further dependence on the considered model system, the direction of the external applied magnetic field (homogeneous or inhomogeneous) is crucial, as it can break the system symmetry and thus generate new solutions. As a result, the classical Navier–Stokes equations become modified to the more complex ferrohydrodynamical equation of motion, incorporating magnetic field and magnetization of the fluid itself, which typically makes numerical simulations expensive and challenging. This chapter provides an overview of the tasks/difficulties from describing and simulating magnetic particles, their interaction, and thus finally resulting modification in rotating flow structures and in particular instabilities and bifurcation behavior.

Keywords: magnetic fluid, complex system, ferrofluid, instabilities, bifurcation, symmetry-breaking, vortex, growth rate, finite-size effect, internal magnetic field, particle interaction, agglomeration, elongational flow

1. Introduction

Ferrofluids [1] represent one example of magnetic particles with a variety of applications, ranging from liquid seals in rotating systems, their use in computer hard drives, as well as in laboratory experiments, to study geophysical flows [2, 3]. However, deeper fundamental study of their magnetohydrodynamics is inevitable, as there are many modeling assumptions that have been implemented to make theoretical descriptions tractable. Ferrofluids are manufactured fluids which consist of dispersions of magnetized nanoparticles in various kinds of liquid carriers. Stabilizing against agglomeration is typically achieved by the addition of a surfactant monolayer onto the magnetic particles. Being no magnetic field present, the magnetic nanoparticles are randomly orientated, which results in a zero-net magnetization of the fluid. Only effect of the present nanoparticles is a typically small alteration to the fluid's density and viscosity. Applying a sufficiently strong magnetic field, the ferrofluid flows toward certain regions of the magnetic field. The hydrodynamics of the system can be significantly changed, and properties of the fluid, e.g., the viscosity, are altered. Therefore, nowadays, they are widely used by various car manufactures in order to provide adaptable suspensions, i.e., changing the damping characteristics of an antishock/damper due to modified viscosity of the ferrofluid under influence of an external applied magnetic field. Both numerical simulations and experiments show that any applied magnetic field, whether radial, axial, azimuthal, transversal, or any combination of the previous stabilizes the basic state [4–9]. This stabilization effect depends, among others, on the particle-particle interactions of the fluid [9]. A strong interaction leads to large/crucial changes in flow behavior, which are driven by the formation of internal flow structures. While an axisymmetric field leaves the flow structures untouched (compared to classical Couette flow), a finite transverse, in-plane magnetic-field component breaks axisymmetry, enforcing new nonlinear effects, modify present flow structures, and can produce new flow structures.

Among various others, one important point to look at is the applied magnetic field, in particular its characteristics while penetrating a moving ferrofluid. In order to describe the hydrodynamics of ferrofluids in containers/boxes typically, models are used which assume that the internal magnetic field and the external applied magnetic field are equal [4–6]. However, such a simplification is only a leading order approximation, as the magnetic field in the container typically is *modified, b*ased on the magnetic susceptibility of the ferrofluid. First work trying to account for modifications [15] includes an expression for the angular velocity of the ferrofluid particles, which differs from the ferrofluid itself due to the magnetic torques acting on the particles. The results are significant *differences* from the external field, providing among others, a much changed body force in the governing equations. More concretely, the interaction between a uniform external field and the susceptibility of the ferrofluid-filled annulus can be shown to result in a field with an $1/r^2$ radial dependence superimposed on the external field.

Further common assumption when describing the hydrodynamics of magnetic particles as ferrofluids is to consider them as point-wise objects and ignore their finite size and specific properties. However, in real, such particles typically aggregate to form clusters having the form of chains, and thus they hinder the free flow of the fluid and increase the viscosity [9–11], and change their rheology and transport properties [14]. In this type of structure's formation, it is also assumed that the interaction parameter is usually greater than unity [1], and thus the strength of the grain-grain interaction can be measured in terms of the total momentum of a particle. Regarding the magneto-dissipative structure of ferrofluid dynamics, one can derive, based on general principles, an equation for macroscopic ferrofluid dynamics. The magnetization's relaxation equation includes an additional term that is proportional to the product of the magnetization's magnitude and the symmetric part of the velocity gradient tensor, which can describe an *elongational flow* scaled by a so-called *transport coefficient* λ_2. Such description also exists in the dynamics of nematic liquid crystals, as the flow alignment modifies the director field in an applied shear flow [16].

It is worth to mention that there are further crucial and important considerations for describing ferrofluids and in general fluids with magnetic particles which are also important as the discussed ones. However, these are not considered in the present chapter. Therefore, we refer to the huge amount of literature, which can be essentially found by the presented references. Two prominent examples of these further considerations are the modulation of the relaxation times of magnetic particles [7, 13], e.g., Neel's or Braun's and the consideration of monodisperse vs. polydisperse materials [7, 17, 18].

In this chapter, we will consider a rotating ferrofluid in Taylor-Couette geometry. First, we are going to describe the modifications in flow solutions and its structural properties and alterations in primary supercritical bifurcation thresholds induced/forced by an external applied magnetic field. Following, we will discuss two different approaches to come to a more realistic description of ferrofluidic flows. Therefore, we will modify and extend the basic ferrohydrodynamical equations. First, considering a modification of the internal magnetic field, which differs from the external applied one due to its internal magnetization. Second incorporating elongational flow. The latter aims to consider finite size effects of the magnetic particles, as well as agglomeration and chain formation procedures.

2. System and theoretical description

As a prototypical system to investigate the influences of a magnetic field on a rotating ferrofluid system, we consider Taylor-Couette flow (Taylor Couette system, TCS) [19, 20] driven by the independent differential rotation of two concentric cylinders (**Figure 1**).

The inner cylinder of radius r_1 rotates at ω_1, and the outer cylinder of radius r_2 rotates at ω_2. If not other stated, in this chapter, periodic boundary conditions are considered in the axial direction with periodicity of length λ and no-slip boundary conditions on the cylinder walls. The system is described using a cylindrical polar coordinate system (r, θ, z) with a velocity field $\vec{u} = (u, v, w)$. The radius ratio of the cylinders is $\beta = r_1/r_2$, and the aspect ratio is $\Gamma = \lambda/(r_2 - r_1)$. The gap between the cylinders is filled with a viscous, incompressible, and isothermal ferrofluid. An external homogeneous magnetic field $\vec{H}_{ext} = H_x \vec{e}_x + H_z \vec{e}_z$ is applied, where $x = r\cos\theta$ is the transverse direction and z is the axial direction, respectively (**Figure 1**).

The flow dynamics of an incompressible homogeneous monodisperse ferrofluid with kinematic viscosity ν and density ρ is governed by the incompressible Navier–Stokes equations, including magnetic terms and the continuity equation. Following, an approach based on the model of Niklas [4, 5] is used, where the magnetic fluid is

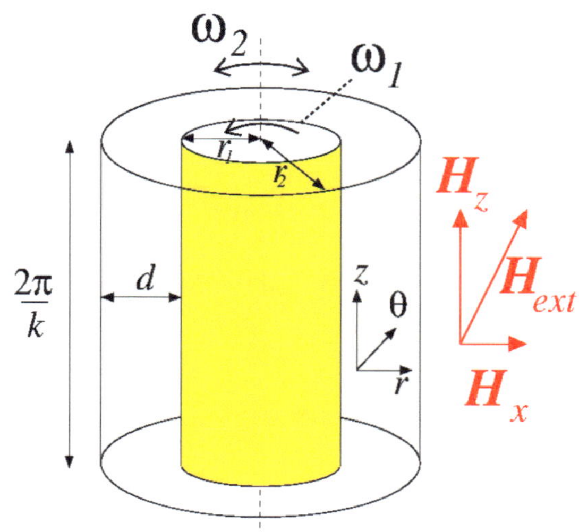

Figure 1
Schematic of the Taylor-Couette system with an external applied homogeneous magnetic field $\vec{H}_{ext} = H_x \vec{e}_x + H_z \vec{e}_z$.

assumed to be incompressible, nonconducting, and to have a constant temperature and a homogeneous distribution of magnetic particles. Using the gap $d = r_2 - r_1$ as the length scale, the diffusion time $\tau_d = d^2/\nu$ as the time scale, scaling pressure with $\rho\nu^2/d^2$, the magnetic field $\vec{H}$, and the magnetization $\vec{M}$ with $(\rho/\mu_0)^{0.5}\nu/d$ (μ_0 is the magnetic constant, i.e., magnetic permeability of free space), the nondimensional governing equations are

$$\left(\partial_t \vec{u} \cdot \nabla\right) \vec{u} - \nabla^2 \vec{u} \, \nabla p = \left(\vec{M} \cdot \nabla\right) \vec{H} + \nabla \times \left(\vec{M} \times \vec{H}\right)/2, \; \nabla \cdot \vec{u} = 0 \tag{1}$$

The cylinders are considered to be of no-slip type with velocity boundary conditions

$$\vec{u}\,(r_1, \theta, z) = (0, \mathrm{Re}_1, 0) \text{ and } \vec{u}\,(r_2, \theta, z) = (0, \mathrm{Re}_2, 0), \tag{2}$$

where the inner and outer Reynolds numbers are $\mathrm{Re}_1 = \omega r_1 d/\nu$ and $\mathrm{Re}_2 = \omega r_2 d/\nu$, respectively.

2.1 Magnetization equations

In order to solve (1), one needs additional information about the magnetic particles themselves, in particular their properties and interaction with an applied magnetic field. Therefore, (1) is solved together with an equation that describes the magnetization of the ferrofluid. A first approximation is to use the equilibrium magnetization of an unperturbed state with a homogeneously magnetized ferrofluid at rest with the mean magnetic moments orientated in the direction of the magnetic field, $\vec{M^{eq}} = \chi\vec{H}$, where χ is the magnetic susceptibility of the ferrofluid, determined using Langevin's formula [21]. The ferrofluid considered for all simulations in this chapter is APG933 [22] with $\chi = 0.9$. However, a ferrofluid's magnetization is also influenced by the flow field. For the presented numerical simulations, an approach based on the model of Niklas [4–6] is used. Furthermore, a stationary magnetization near equilibrium considering a small perturbation at the equilibrium magnetization $\left(\left(\vec{M} - \vec{M^{eq}}\right)\right)$ and small relaxation times $\Omega\tau << 1$, where Ω is the absolute value of half of the vorticity Ω and τ is the magnetic relaxation time. In the near-equilibrium approximation, Niklas [4] determined the relationship between the magnetization $\vec{M}$, the magnetic field $\vec{H}$, and the velocity $\vec{u}$ to be

$$\vec{M} - M^{eq} = c_N \, \vec{\Omega} \times \vec{H} \tag{3}$$

with $\vec{\Omega} = \nabla \times \vec{u}\,/2$ being the vorticity and the *Niklas coefficient*

$$c_N^2 = \tau / \left(1/\chi + \tau\mu_0 H^2/(6\mu\Phi)\right) \tag{4}$$

where μ is the dynamic viscosity and Φ is the volume fraction of the magnetic material.

2.2 Ferrohydrodynamic equations of motions

The magnetization can be eliminated from (1) by using (3) in order to obtain the ferrohydrodynamic equation of motion [4]:

$$\left(\partial_t + \vec{u}\cdot\nabla\right)\vec{u} - \nabla^2\vec{u} + \nabla p_M = -c_N^2\left(\vec{H}\,\nabla\cdot\vec{F} + \vec{H}\times\nabla\times\vec{F}\right)/2 \qquad (5)$$

where $\vec{F} = \vec{\Omega}\times\vec{H}$ and p_M are the dynamic pressure incorporating all magnetic terms which can be written as gradients. Note that the relaxation time τ could typically be defined by $\tau = \tau_B\langle D^3\rangle^{1/3}$, where τ_B is the Brownian relaxation time and $\langle D^3\rangle^{1/3}$ is the averaged diameter of the magnetic particles. For simplicity, to investigate the different effects, i.e., internal magnetization and elongational flow (see below), we assume that τ is constant, which is independent of the magnetic field $\vec{H}$. The considered relaxation time is based on the experimental result [22] $(\tau = \tau_{APG933}/\tau_d = 0.0018)$.

Consider external field to be equal to internal one $\left(\vec{H} = \vec{H}_{ext}\right)$, (5) can be simplified to

$$\left(\partial_t + \vec{u}\cdot\nabla\right)\vec{u} - \nabla^2\vec{u} + \nabla p_M = -c_N^2\left(\vec{H}\,\nabla\cdot\vec{F} + \vec{H}\times\nabla\times\vec{F}\right)/2 \qquad (6)$$

In this approach, the magnetic field and all the magnetic properties of the ferrofluid influence the velocity field only via the magnetic field, the *Niklas parameter* $\vec{s_N} = s_x\vec{e_x} + s_z\vec{e_z}$ [4–6], whereby magnetic field strengths are:

$$s_x = \left[2(2+\chi)/\left((2+\chi)^2 - \chi^2\eta^2\right)\right]H_x c_N \text{ and } s_z = H_z c_n \qquad (7)$$

In this approach, s_x and s_z are dimensionless numbers, describing the strength of the applied magnetic field.

3. Ferrofluidic flow under finite magnetic field

3.1 Introduction and motivation

We start our discussion with the investigation of the influence of a homogeneous magnetic field on a ferrofluid. In this first part, we will assume that the internal magnetic field is equal to the external applied magnetic field, i.e., homogeneous. Moreover, we ignore in this section any kind of finite size effects, e.g., possible agglomeration and chain formation, later considered as elongational flow effect (Section 4). Therefore, we consider the ferrofluid particles as point-size objects without any 3D expansion. Structure, dynamics, symmetry properties, bifurcation, and stability behavior of different vortex structures are investigated for axial and transversal magnetic fields, as well as combinations of them, i.e., oblique magnetic field. We will see that a transversal magnetic field has significant influence. It modulates and changes the classical flow structures, as well the basic state (Circular Couette flow, CCF) as the classical primary supercritical ones, Taylor vortex flow (TVF) and the spiral vortex flow (spirals, SPI). They are replaced by wavy-modulated flow structures: wavy Taylor vortex flow (wTVF) and wavy spiral vortex flow (wSPI), as well as they forward bifurcate out of the modified 2-fold circular Couette flow (2-CCF).

3.2 Flow structures

3.2.1 Flow structures in absence of a magnetic field

There are two fundamental basic, topologically different flow structures appearing in (classical) Taylor-Couette flow (see also Nomenclature and

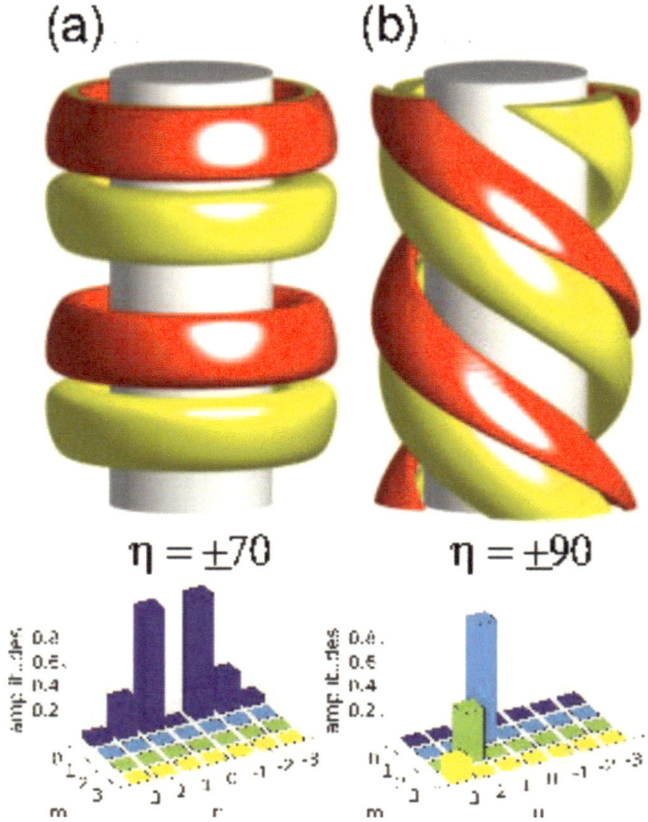

Figure 2.
Azimuthal vorticity isosurfaces η over two axial wavelengths (for visual purpose) in absence of a magnetic field (N) of (a) TVF at $\mathrm{Re}_2 = 0$, $\mathrm{Re}_1 = 100$, and (b) SPI at $\mathrm{Re}_2 = -150$, $\mathrm{Re}_1 = 150$. Red (yellow) isosurfaces correspond to positive (negative) values as indicated. The bottom row shows mode amplitudes $|u_{m,n}|$ of the radial velocity field u corresponding to structures above over the m-n plane. The values are scaled regarding the maximum mode amplitude to be 1 (TVF: $|u_{0,1}| = 1$).

abbreviations section). These are either toroidally closed Taylor vortices (Taylor vortex flow, TVF) and helical spiral vortices (spiral vortex flow/spirals, SPI) [20]. **Figure 2** shows the structure of these fundamental solutions, TVF and SPI (here for a left-handed SPI with azimuthal wavenumber $m = 1$), by using isosurfaces of the azimuthal vorticity. Because of the system symmetry $z \to -z$, we only discuss left-handed SPI. Without other parameters, e.g., axial through flow, left- and right-handed SPI are mirror images of each other. These pure states can also be found modulated as wavy structures, which are topologically identical with the pure states. However, they show a wavy-like deformation that results from additional and other stimulated modes compared to the classical solutions.

While *classical* wavy flow solutions [27–36] are well known and have already been widely discussed in literature regarding TCS, the ones generated by a magnetic field are different. In particular, the magnetically induced wTVF are *nonrotating* structures having a *pinned* phase, contrary to the classic wTVF, which *rotate* azimuthally. Regarding the flow structure, the imposed magnetic field shrinks and expands the vortices at different cross sections of constant θ, resulting in an azimuthal alternating wide and narrow vortex belly structure.

3.2.2 Flow structures—mode coupling in magnetic fields

Following, we will have a quick look at the impact on the different applied magnetic field with respect to the flow structures and in particular their mode

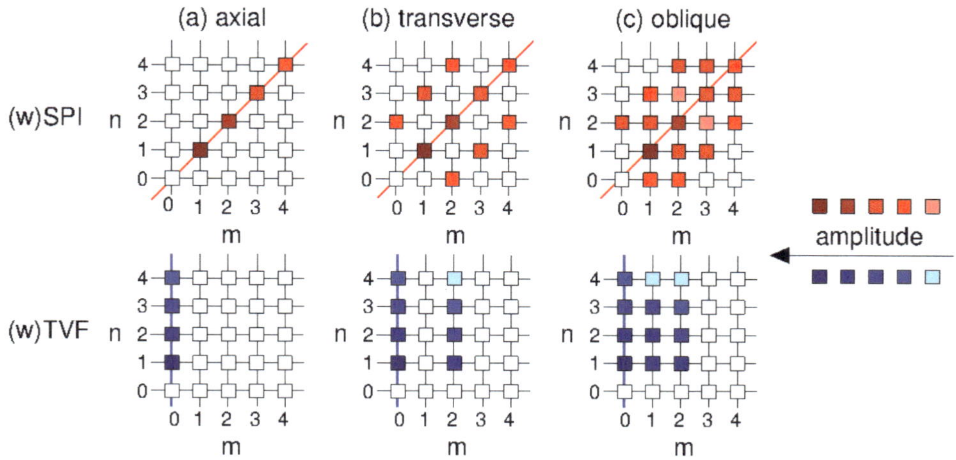

Figure 3.
Schematic sketch of modes in the presence of different externally imposed (homogeneous) magnetic fields. Here, m represents the azimuthal, and n the axial mode index in the Fourier expansion (Appendix (17) and (18)). (a) Modes either in absence of any magnetic field or in a pure axial magnetic field. (b) Pure transversal magnetic field with additional stimulation of $m = n \pm 2$ modes. (c) Superposition of axial and transversal fields with even further finite $m = n \pm 1$ modes. Strength of mode amplitudes, the magnitudes are characterized from dark (large amplitudes) to bright (small amplitudes) colors (legend right). Red and blue lines indicate the dominant mode amplitudes of the flow structures.

spectra. The modifications in NSE due to the additional terms incorporating magnetization and magnetic field result in various new phenomena. Most important is probably the fact that magnetic fields with a finite transversal component stimulate additional modes. Depending on different external applied magnetic fields, **Figure 3** presents a schematic plot for the different stimulated modes, depending on the applied field.

Having a *pure* axial field ($s_x = 0, s_z > 0$) applied the mode spectra, the flow structures remain qualitatively similar to the no field (N) case (**Figure 3(a)**). A TVF still contains only the axisymmetric modes $m = 0$ (with dominant n = 1, see also **Figures 2** and **6**). Similarly, L1-SPI, i.e., a left-handed SPI flow with azimuthal wavenumber $m = 1$ contains modes only on the diagonal $m = n$ (with dominant $n = 1$, see also **Figures 2** and **6**). However, if a field with finite transversal component is applied additional mode become finite and the flow structures change. A pure transversal field ($s_x > 0, s_z = 0$) excites $m = \pm 2$ modes lying on the secondary diagonal $m = n \pm 2$ (**Figure 3(b)**). Thus, as a result, the pure states of TVF and SPI vanish in the presence of a transversal magnetic field. They become wavy-like modified, resulting in wTVF and wSPI. Finally, having a magnetic field with either axial and transversal component ($s_x > 0, s_z > 0$), there are further interactions, which result in stimulation of additional $m = n \pm 1$ modes (**Figure 3(c)**). Moreover, these additional modes can induce further nonlinear mode couplings. As a result, the mode spectra contain more nonzero components (**Figure 3(c)** and **Figure 6** for numerical results). Thus all flow structures in a magnetic field with transversal component differ qualitatively from the classical flows found in the absence of magnetic fields [6, 26].

3.3 Effect of finite magnetic fields

3.3.1 Basic state and symmetries

If only a pure axial magnetic field, $H_{ext}(r)\vec{e_z}$ (see **Figure 1**), the classical circular Couette flow (CCF) is observed, with $\vec{u}_{CCF} = (0, Ar + B/r, 0)$, where

$A = (\mathrm{Re}_2 - \beta Re_1)/(1+\beta)$, and $B = \beta(\mathrm{Re}_1 - \beta Re_2)/(1+\beta)(1-\beta)^2$, is a solution of (5) as the associated vorticity $\vec{\Omega}_{ccf} = \nabla\times\vec{u}_{ccf}$ is parallel to the applied magnetic field and therefore all magnetic terms vanish. Although for magnetic fields that are orientated purely in the radial or azimuthal direction, the basic state changes but remains axisymmetric with deviations from CCF only having an azimuthal component [4, 23]. As a result, in all of these cases, the basic state is invariant to a number of symmetries (azimuthal rotation, mirror symmetry, and axial translation) whose actions on a general velocity field are

$$R_\Phi(u,v,w)(r,\theta,z,t) = (u,v,w)(r,\theta+\Phi,z,t),\ \Phi\in[0,2\pi], \quad (8)$$

$$K_z(u,v,w)(r,\theta,z,t) = (u,v,-w)(r,\theta,-z,t), \quad (9)$$

$$T_\alpha(u,v,w)(r,\theta,z,t) = (u,v,w)(r,\theta,z+\alpha\Gamma,t),\ \alpha\in N \quad (10)$$

On the contrary, for a field with a finite transverse component, the base state is no longer axisymmetric. Although the basic state remains invariant to T_α and a modified mirror symmetry K_z^H [24], a transverse magnetic field T breaks the continuous axisymmetry R_Φ, resulting in a basic state with discrete symmetry R_π, i.e., with azimuthal wavenumber $m = 2$ [24]. Following, we will refer to this modified 2-fold symmetric CCF as 2-CCF. It is important to mention that this $m = 2$ state is stationary [24], which is in contrast to the generic breaking of a $SO(2)$ symmetry which would result in a rotating wave.

3.3.2 Bifurcation behavior

As seen in Section 3.2, an applied magnetic field can stimulate additional modes and thus changes the Fourier mode spectrum of flow structures resulting in new solutions. The results are wavy-like modified flow states. It is to expect that these mode-stimulated modifications also influence and appear in the primary bifurcation thresholds of the solutions, as well as in their growth rate. Following, we will have a look at the bifurcation behavior of the different flow structures with the additionally field-induced modes explained in Section 3.3. In particular, we will focus on four different field configurations: *no field (N)* ($s_x = 0, s_z = 0$); *pure axial field (A)* ($s_x = 0, s_z = 0.6$); *pure transversal field (T)* ($s_x = 0.6, s_z = 0$); and *oblique field (O)* ($s_x = 0.6, s_z = 0.6$) ((7) for parameters s_x, s_z).

Considering these four, non, pure axial, pure transversal, and oblique magnetic fields, **Figure 4** illustrates the stable forward bifurcating branches of toroidally closed TVF and of wTVF solutions, respectively. **Figure 5** shows the bifurcation diagrams for helical SPI and wSPI in an analogous way. Note that the shown

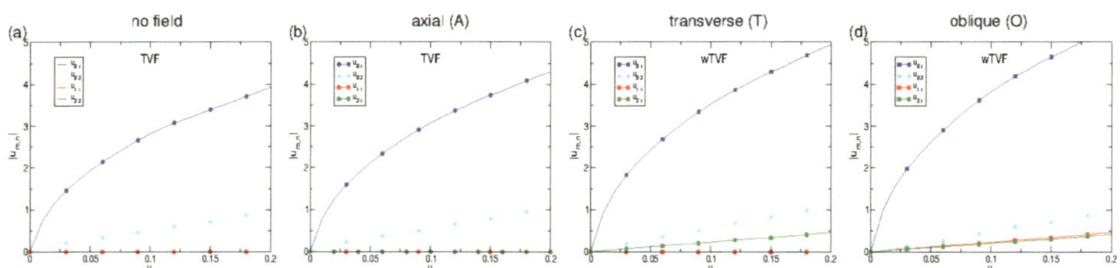

Figure 4.
Bifurcation diagrams of TVF and wTVF, respectively, in different magnetic fields: (a) no field (N), (b) axial (A), (c) transverse (T), and (d) oblique (O). Moduli $|u_{m,n}|$ *(18) of the radial flow field amplitudes at mid gap are shown versus the relative distance* $\mu = \mathrm{Re}_1/\mathrm{Re}_{1,c} - 1$ *from the onset of the respective vortex flow. Further parameters are:* $\mathrm{Re}_2 = 0, \beta = 0.5, k = 3.927$ *(cf.* ***Figure 2****).*

wavy-like modulated flow states only result from symmetry breaking due to finite transversal field component. In the **Figures 4** and **5**, the relative distance $\mu = \mathrm{Re}_1/\mathrm{Re}_{1,c} - 1$ is used as control parameter for easier comparison of the bifurcation branches ($\mathrm{Re}_{1,c}$ stands for the critical onset Reynolds number Re_1 of the flow structure, which depends on various system parameters. These are the aspect ratio Γ, the radius ratio r_1/r_2, the outer Reynolds number Re_2, axial boundary conditions, and the field strength s_x, s_z, respectively). The main reason to use μ is the shift in the bifurcation onset, which varies depending on the applied field direction.

Displayed in the **Figures 4** and **5** are the moduli $|u_{m,n}|$ of the dominant, as well as the first higher harmonic modes of the corresponding structure (see Appendix (18) and (19)), e.g. $(m,n) = (0,1)$ and $(m,n) = (0,2)$ for TVF. Furthermore, the largest field-induced mode amplitudes $|u_{2,1}|$ and $|u_{1,\pm 1}|$ (only in oblique configuration) of the wavy structures (cf. schematics in **Figure 3**) are included for comparison.

Axial field (A): As a pure axial magnetic field ($s_x = 0, s_z \neq 0$) does not stimulate any further modes within the mode-spectra (**Figure 3(a)**), the flow states TVF and SPI and their structural properties remain qualitatively similar, and they are as without a field. However, quantitative, they eventually change. First important effect is the rise of the onsets to higher values of Re_1 [6, 7]. Thereby, the upward shift depends on the various system control parameters (e.g., $\mathrm{Re}_2, \beta, \Gamma$). For parameters in **Figure 4**, the onset of TVF rises by about 25%, and for the parameters of **Figure 5**, the SPI bifurcation threshold moves upward in by about 15%. For the latter, consequently also corresponding SPI frequencies are increased in an axial field. Other change regarding **Figure 4** is the amplitude of the modes with increasing the parameter μ (weakest for nonfield and strongest for oblique field configuration).

Transversal field (T): A transversal magnetic field ($s_x \neq 0, s_z = 0$) stimulates additional azimuthal modes $m = n \pm 2$ (Section 3.2) as illustrated in **Figure 3(b)**. Thus, based on this knowledge, one can already expect a more dramatic influence on the classical TVF and SPI structures. As a pure axial field, a transversal field also delays the onset of the primary bifurcating structures, i.e., it stabilizes the basic state. Furthermore and more important is the fact that such a field also changes the flow structural properties in qualitative manner. The growth of the additional modulus $|u_{2,1}|$ (18) (for toroidally closed flow states) with $\mu = \mathrm{Re}_1/\mathrm{Re}_{1,onset} - 1$ changes the pure TVF to a wavy-like modulated wTVF. Analog pure SPI changes to wSPI (shown for L1-(w)SPI in **Figure 5**) due to the additional $m = n \pm 2$ (see **Figure 3(b)**), indicated by the growth of the modulus $|u_{3,1}|$ in **Figure 5**. It is worth to mention that due to the mode stimulation and interaction, also the (1,-1) mode becomes excited, which is similar to structure R1-(w)SPI. Note, the same also holds for the symmetry-related mirror image L1-(w)SPI.

Thus, in the presence of a magnetic field with a finite transversal component, the *pure/classical* TVF and SPI structures are modified. They are replaced by their

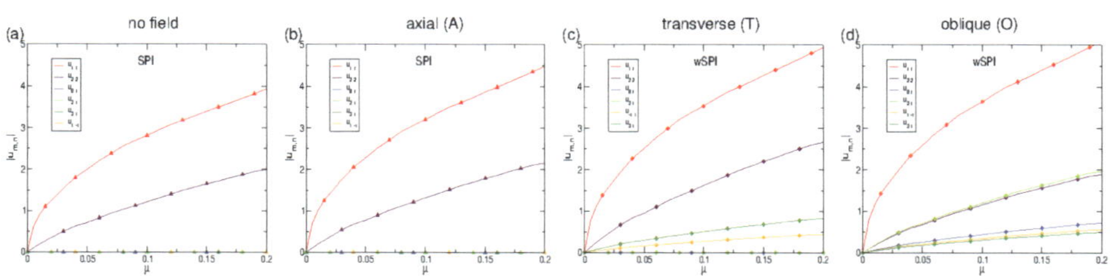

Figure 5.
Bifurcation diagrams of SPI and wSPI, respectively, in different magnetic fields: (a) no field (N), (b) axial (A), (c) transverse (T), and (d) oblique (O). Moduli $|u_{m,n}|$ *(18) of the radial flow field amplitudes at mid gap are shown versus the relative distance* $\mu = \mathrm{Re}_1/\mathrm{Re}_{1,onset} - 1$ *from the onset of the respective vortex flow. Further parameters are:* $\mathrm{Re}_2 = -150, \beta = 0.5, k = 3.927$ *(cf.* ***Figure 2****).*

wavy-like counterparts, wTVF and wSPI, which now bifurcate as *primary* structured solutions forward out of the CCF-2 ground state; note that already the basic state has an additional mode-2 symmetry due to the symmetry-breaking effect of the transversal field and as such already differ from the classical CCF (Section 3.3.1). As for axial field, the magnitude of stabilization varies with given system control parameters. In general, it is slightly smaller compared to the stabilization effect of an axial field. For instance for parameters presented in **Figures 4** and **5**, the upward shift for the onset of wTVF (wSPI) is about 20% (12%) in an axial (transversal) magnetic field.

Important to mention that a finite transverse component $s_x \neq 0$ breaks the system symmetry but does not change the degeneration between left- and right-handed SPIs. Both solutions keep bifurcating at the same critical values. The latter is different to the effect of an axial flow, for instance. Aside that such a flow also breaks the system symmetry, it separates both left- and right-handed SPI solutions. Under axial applied flow, they appear at different critical values.

Oblique field (O): As already seen for both pure field configurations, either $s_x \neq 0, s_z = 0$ or $s_x = 0, s_z \neq 0$, the onset of (w)TVF and (w)SPI is shifted to higher values. In fact in a field, obliquely orientated to the cylinder ($s_x \neq 0, s_z \neq 0$), the magnitude of shift is bigger compared to the pure cases. In addition, as discussed before, such a superposition of an axial and a transversal field stimulates also new, additional modes with higher mode index combinations $m = n \pm 1$ (**Figure 3(c)** and for numerical simulations **Figures 4(c,d)**, **5(c,d)**, and **6(e,f)**).

A common finding for (w)TVF and (w)SPI is that their dominant mode amplitudes for $m = 0$ (for (w)TVF) and $m = 1$ (for (w)SPI) grow with respect to the bifurcation onsets. Although the absolute value depends on the different applied magnetic field, comparing the moduli $|u_{m,1}|$ at the same relative distance μ from the respective onset (**Figures 4** and **5**, compare dominant mode amplitudes), one finds the following relations

$$|u_{m,1}(s_x = c, s_z = c)| \geq |u_{m,1}(s_x = c, s_z = 0)| \geq |u_{m,1}(s_x = 0, s_z = c)| \geq |u_{m,1}(s_x = 0, s_z = 0)|$$

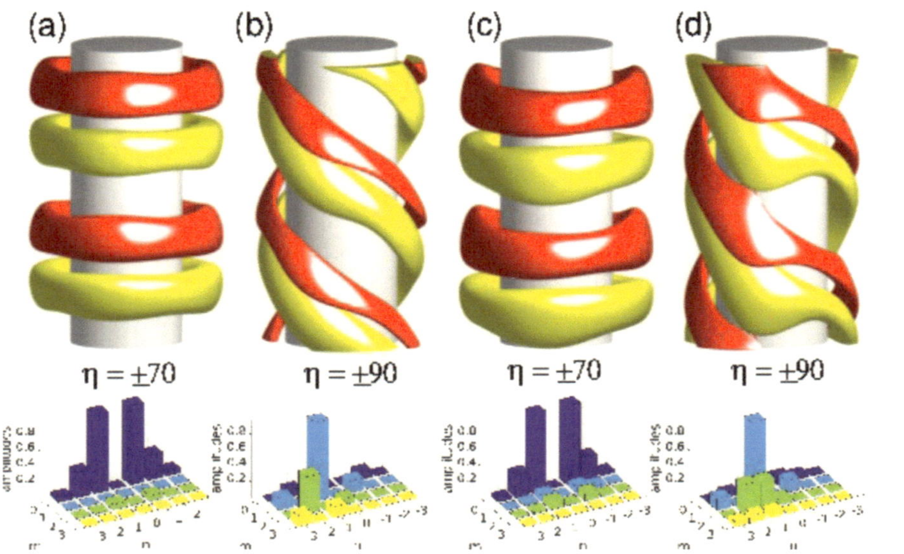

Figure 6.
Azimuthal vorticity isosurfaces η over two axial wavelengths (for visual purpose) with pure transversal applied magnetic field (T): (a) wTVF at $Re_2 = 0$, $Re_1 = 125$, *(b) wSPI at* $Re_2 = -150$, $Re_1 = 160$; *and oblique applied magnetic field (O): (c) wTVF at* $Re_2 = 0$, $Re_1 = 150$, *and (d) wSPI at* $Re_2 = -150$, $Re_1 = 190$. *Red (yellow) isosurfaces correspond to positive (negative) values as indicated. Bottom row shows mode amplitudes* $|u_{m,n}|$ *of the radial velocity field u corresponding to structures above over the m-n-plane. Values are scaled with the maximum mode amplitude to be 1 (TVF:* $|u_{0,1}| = 1$*).*

for a given constant value $c\in[0,1]$ and $m = 0$ in case of toroidally closed (w)TVF and $m = 1$ for helical (w)SPI, respectively (Compare upper curves for dominant mode amplitudes $|u_{m,1}|$ in **Figures 4** and 5).

For helical flow structures, the larger variety of mode interactions due to the basic spectra (e.g. $m = n$ $[m = -n]$ for pure L1-SPI [R-SPI]) results in significant variation of the higher order and secondary stimulated modes, depending on the field configuration (see in particular **Figures 3(c)**, **4(d)**, and **5(d)**).

Based on the both bifurcation diagrams presented in **Figures 4** and **5**, it is worth to remind that in the case of a finite transversal field component $s_x \neq 0$, no pure solutions, TVF and SPI, exist. They become replaced by wavy-like modulated wTVF and wSPI. However, independent of the field, leading mode amplitude of the respective flow state, (w)TVF and (w)SPI, follows a typical square-root forward bifurcation, whereby the slope of its square grows with the applied field strength.

3.3.3 Vortex structures and mode contents

Having discussed the changes that different oriented magnetic fields can cause in the bifurcation behavior and mode stimulation of (w)TVF and (w)SPI, respectively, one can imagine that these also coincide with flow structural modifications. Now, we will address these structural changes of the vortices and the associated changes in the mode contents of the flows in more detail.

Figure 6 shows isosurfaces of the azimuthal vorticity η over two axial wavelengths (for visual purpose) for supercritical flow states (w)TVF and (w)SPI in the three different considered field configurations: (a,b) no field applied (N), (c,d) pure transverse field (T), and (e,f) oblique field (O) at parameters Re_1 and Re_2 as indicated.

A good quantity in order to visualize the field induced changes in the 3D vortex structures are isosurfaces of the azimuthal vorticity η. These surfaces appropriately convey structural details of the vortex flows in question [28].

Axial field (A): As a pure axial field does not break any continuous symmetry of the system, there is neither qualitative change in the structure of TVF and SPI in real space nor the mode structure of the flow in the $m - n$ Fourier plane (not shown, but similar as **Figure 2(a,b)**; only quantitative the mode amplitudes are slightly larger). TVF still consists of toroidally closed, rotationally symmetric vortex tubes with a mode spectrum containing only $m = 0$ modes (**Figure 2(a)**). And also, the helically oriented, open vortex tubes of SPI (here for left-winding spiral L-SPI) containing $m = n$ modes as shown in **Figures 2(b)** and **3** (mirror image of right-winding R-SPI analog) are *unaffected* by an axial field.

Transversal and oblique field (T, O): Having a magnetic field with finite transversal component crucially changes the scenario as TVF, and SPI structures are replaced by modified flow states due to the presence of such a magnetic field. In general, the isosurfaces become wavy-like deformed, and additional Fourier modes are stimulated as described before (Section 3.3; see also schematics in **Figure 3**). For pure transversal field, this mode stimulation results in wTVF with additional $m = \pm 2$ modes aside the $m = 0$ (**Figure 6(c)**) and in wSPI with $m = n \pm 2$ modes on the secondary diagonal(s) in addition to the $m = n$ SPI modes on the diagonal (**Figure 6(d)**). Beyond these, an oblique field further stimulates the modes $m = \pm 1$ in wTVF and $m = n \pm 1$ in wSPI due to more complex nonlinear dynamics/mode interaction. In particular, the spatio-temporal properties of the wTVF generated due to symmetry breaking magnetic field, either transversal or oblique (**Figure 6(c,e)**), differ significant from those of the classic wTVF. The structure of classical wTVF *rotates* as a whole in azimuthal direction [27, 32–35].

The wavy-like deformation on the toroidally closed vortices η introduced by a magnetic field with a finite transversal component is different. Regarding **Figure 6(c,e)**, one sees that the tubes of isosurfaces of the azimuthal vorticity become strong modified and bent at certain azimuthal cross sections of constant θ in comparison with the pure TVF tubes. Classic wTVF arises by deforming the TVF such that the axial stack of closed vortex tubes is wave-like modulated, bending axially upward and downward. Then, this deformation structure of the vortices rotates as a whole with a characteristic frequency. In contrast to these classical rotating wTVF, the wTVF initiated due to the symmetry-breaking transversal field component is *stationary* and does *not rotate* in azimuthal direction, instead, developing an 2-fold belly structure. It is worth to mention that such 2-fold symmetry also applies to the basic state. Thus, the classical rotational, axisymmetric CCF is replaced by the 2-CCF due to the symmetry breaking in the transversal magnetic field (see Section 3.2).

Consider wSPI, one can identify qualitative similar wavy characteristics, whereby the topological differences are weaker in comparison to wTVF. As the classical SPI structures already rotate without magnetic field, this also remains for wSPI in the presence of magnetic fields. However, also wSPI remains structural almost identical, and their modulation amplitude becomes enforced at certain cross sections of constant θ in presence of a transversal field component.

The basic state (CCF) loses stability either to (w)TVF (**Figure 5**) or (w)SPI (**Figure 6**). In case of wTVF, a supercritical pitchfork of revolution bifurcation breaks the axial translational symmetry T_α (8.3) and parameterizes the axial location of the steady axisymmetric family of solutions. On the other hand, (w)SPIs appear in a supercritical Hopf bifurcation which breaks the mirror symmetry K_z (8.2), resulting in a pair of symmetrically related spiral states, one with left-handed and the other with right-handed winding. It is worth to mention that a finite transverse magnetic field T breaks the axisymmetry R_φ, resulting in a basic state with discrete symmetry R_π, i.e., with azimuthal wave number $m = 2$ (section 3.3.1). Note that this $m = 2$ state is *stationary,* which is in contrast to a generic breaking of an $SO(2)$ symmetry, resulting in a rotating wave.

3.3.4 Stabilization of basic state: point of higher co-dimension

For TCS, a crucial point of special interest in the $\mathrm{Re}_1 - \mathrm{Re}_2$ phase diagram is the bicritical, co-dimension two point, γ, at which the primary bifurcation thresholds of TVF and SPI intersect/converge and as thus exchange the stability of the primary bifurcating solution [20]. For instance, in absence of a magnetic field and given wavenumber $k = 3.927$, the γ-point is given by $\mathrm{Re}_1 \approx 95.25$ and $\mathrm{Re}_2 \approx -73.69$ [28] (**Figures 7(a)** and **8(a)**); for smaller Re_2 (stronger counter-rotation), SPI bifurcate *stable,* and for larger Re_2 (less counter-rotation and co-rotation), TVF bifurcate stable respectively with increasing Re_1 (**Figure 8(a)**). Thus, with increasing Re_1, the 2nd solution bifurcates as *unstable* solution. Note that the γ-point is quite sensitive to the axial wavenumber; for different k, it moves toward higher Re_1 and smaller Re_2. This bifurcation scenario also holds for wTVF and wSPI in oblique and transversal magnetic fields.

It is worth to mention that the curves in **Figures 7(a)** and **8** present "*marginal stability*" thresholds. These are obtained for a fixed wavenumber $k = 3.927$, which differ from the '*critical*' thresholds. The latter is defined by the (smallest) Reynolds number, for which the flow structure first appears, which crucially depends on k. Thus, the intersection points of the curves for TVF and SPI (e.g., **Figure 7(a)**) are critical values, with respect to the fixed wavenumber.

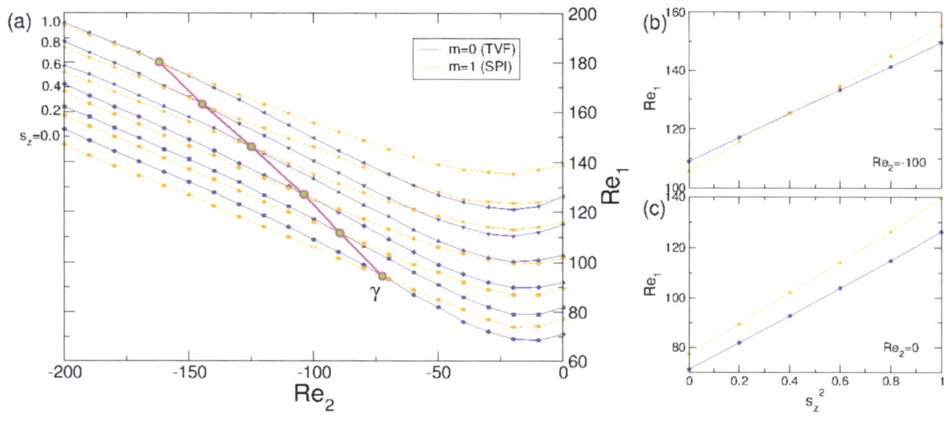

Figure 7.
(a) Influence of an axial magnetic field on the location of the bifurcation thresholds of TVF and SPI out of CCF in the $\mathrm{Re}_1 - \mathrm{Re}_2$ *diagram. Blue (full) lines refer to TVF and orange (dashed) ones to SPI. Magenta line with points indicates the moving of the* γ*-point with* s_z *towards smaller* Re_2*. Further shown are the bifurcation thresholds* $\mathrm{Re}_{1,c}$ *with increasing* s_z^2 *for fixed (b)* $\mathrm{Re}_2 = -100$ *and (c)* $\mathrm{Re}_2 = 0$*, respectively, obtained by full non-linear simulations.*

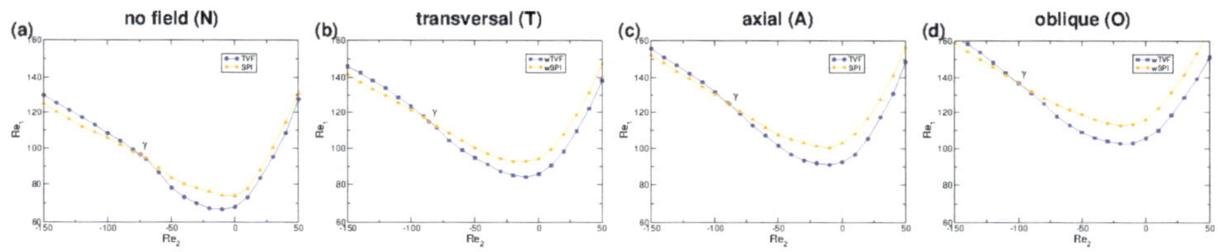

Figure 8.
Bifurcation curves for the onset of (w)TVF (blue) and (w)SPI (orange) in $\mathrm{Re}_2 - \mathrm{Re}_1$ *diagram under (a) no magnetic field (N)* $(s_x = 0, s_z = 0)$*, (b) an applied transverse field (T)* $(s_x \neq 0, s_z = 0)$*, (c) an applied axial field (a)* $(s_x = 0, s_z \neq 0)$*, and (d) an applied oblique field (O)* $(s_x \neq 0, s_z \neq 0)$*. Simulation results of full nonlinear Eqs. (11) consider short periodic TCS with* $k = 3.927$ *(cf. [6, 28]).*

Figure 7(a) illustrates an example for the influence of different axial magnetic field strength s_z on the bifurcation thresholds for TVF and SPI and the variation on the location of this bicritical γ point. These are results of linear calculations and were obtained with a shooting method [7]. Solid blue (dashed orange) lines refer to TVF (SPI). Note that always the lower (in $\mathrm{Re}_{1,c}$) bifurcating vortex solution is stable and the upper one is unstable. As discussed in Section 3.2, increasing s_z stabilizes the CCF basic state, both, against TVF and SPI, i.e., the upward shift of the respective threshold curves. In general, the strength of the stabilization effect is larger for helical SPI vortex structures than for toroidally closed TVF (e.g. **Figure 8**). As a result, the γ-point moves toward more negative Re_2 (i.e., toward the left in **Figure 7(a)**). Thus, it is observed that the Re_2 region of primary, stable bifurcating TVF expands with increasing s_z.

The same scenario also applies for transversal and oblique magnetic field configuration, whereby the magnitude of the shift is biggest for a pure axial magnetic field and almost equal for a pure transversal and oblique magnetic field (slightly larger for the latter). **Figure 7(b,c)** displays the bifurcation thresholds $\mathrm{Re}_{1,c}$ for fixed outer cylinder rotations, $\mathrm{Re}_2 = -100$ and $\mathrm{Re}_2 = 0$ with increasing s_z^2, resulting from fully nonlinear simulations. The slopes in **Figure 7(b,c)** depend crucially on Re_2. However, in general, the slopes for SPI are steeper than for TVF. For small to moderate field parameters, TVF is the primary stable bifurcating structure while

SPI bifurcate secondarily and unstable. For stronger fields, the structures interchange stability and bifurcation order.

Summarizing, for fixed Re_2, an axial field shifts the onset of helical SPI more strongly than the one of toroidally closed TVF. We furthermore found this behavior to appear for all parameters and magnetic fields: either for axial, transversal, and oblique magnetic fields (**Figure 8**). In this case, that for transversal and oblique fields, all structures are wavy-like modulated; wTVF and wSPI, respectively (see also Section 3.2).

In pure axial or pure magnetic fields, the upward shifts of the bifurcation values, the stabilization of the basic state (CCF or 2-CCF), i.e., the onset of the primary stable flow state (w)TVF or (w)SPI grows linearly (**Figure 7(b,c)**) with the squared magnetic field parameters s_x^2 or s_z^2, respectively [6]. Thereby, the magnitude of stabilization shift depends crucially on Re_2, in general is largest close to outer cylinder at rest.

3.3.5 Marginal stability thresholds: Stabilization of the basic state

Before going into detail about the stabilization of the basic state it is worth, once more to mention, that the basic state can be different in presence of a magnetic field, whether the field contains a symmetry breaking transversal component or not. For pure axial field, one finds the classical CCF which becomes replaced by the 2-fold symmetric 2-CCF for fields with finite transversal component (see also Section 3.3.1).

Figure 8 presents bifurcation thresholds for vortex flows under the influence of different magnetic fields. Shown is the $Re_2 - Re_1$ diagram, along which the CCF state loses stability. For comparison, (a) shows the thresholds in absence of any field ($s_x = 0, s_z = 0$), while the same thresholds are shown for (b) a transversal ($s_x \neq 0, s_z = 0$), (c) an axial ($s_x = 0, s_z \neq 0$), and (d) an oblique field ($s_x \neq 0$, $s_z \neq 0$). These results were obtained by nonlinear simulations with the full field ferrohydrodynamical Eqs. (11), considering a short periodic TCS with $k = 3.927$. See also **Figure 7**, for linear stability analysis, in which the shift of the higher co-dimension γ-point is shown together with stabilization of CCF in an axial field.

For all bifurcation thresholds of TVF, SPI, wTVF, and wSPI upward shifts of the bifurcation values, $Re_{1,c}$, as shown in **Figure 8**, are observed. However, the exact onsets of the different primary bifurcating structures differ, either depending on the particular applied field as well as on Re_2.

In absence of a magnetic field (**Figure 8(a)**) and sufficiently strong counter rotation, $Re_2 \leq -73$ [28], the basic state, CCF, first becomes unstable to a helical solution, SPI, in a supercritical Hopf bifurcation breaking K_z, which results in a pair of symmetrically related, degenerated spiral states, one with left-handed winding and the other with right-handed winding. For considered Re_2-parameter range in **Figure 8**, the bifurcating SPI has an azimuthal wavenumber $m = 1$ (or $m = -1$ in case of the right-handed SPI). Note that the wavenumber m of the primary bifurcating SPI increases with stronger counter-rotating cylinders [20]. On the other hand, for $Re_2 \geq -73$, the basic state loses stability to the toroidally closed TVF. The supercritical pitchfork of revolution bifurcation breaks T_α, and the steady axisymmetric family of solutions are parametrized by their axial location.

Considering a pure axial magnetic field, the scenario remains mainly unchanged (**Figure 8(c)**); still SPI and TVF appear as primary supercritical flow states with an intersection of the bifurcation thresholds at a specific value Re_2. As already seen in **Figure 7**, increasing field strength s_z in a pure axial field tends to shift the co-dimension γ-point toward more negative Re_2. The same also holds for increasing s_x in a pure transversal field, with significant minor effect on Re_2 variation. Thus, the magnetic fields tend to render the corresponding basic states more unstable to (w)TVF and more stable to (w)SPI, respectively.

Having a field with finite transversal component, either pure transversal (**Figure 8(b)**) or oblique orientated (**Figure 8(d)**), the primary instabilities, with TVF being replaced by wTVF, and SPI being replaced by wSPI, respectively, just as the basic states were modified from CCF to 2-CCF due to the presence of the magnetic field. Their onset $Re_{1,c}$ occurs at higher Re_1 values for any given Re_2 value compared to the onset of classical TVF and SPI in the absence of a magnetic field. Thus, the magnetic fields alter the basic states and make them more robust to instabilities. The level of stabilization is greater for a pure axial field compared to a pure transversal field, but strongest, when an oblique field is imposed (bifurcation values $Re_{1,c}$ grow with field parameters s_x, s_z).

An oblique magnetic field results in the strongest upward shift of bifurcation values $Re_{1,c}$, and also, this shift depends on Re_2, which interestingly is largest about $Re_2 = 0$, outer cylinder at rest. Furthermore, the topological different flow structures of (w)TVF and (w)SPI are shifted about the same amplitude, with a slightly lager effect on helical (w)SPI for the considered $Re_2 - Re_1$ diagram. This should be kept in mind regarding later consideration involving field modulation via internal dependence, presented in Section 5.

So, while a pure axial fields shift the bifurcation onsets more strongly (i.e. to larger values Re_1) than pure transversal ones (**Figure 8**), the latter one has a stronger influence on the nonlinear spatio-temporal properties and dynamics of the bifurcating vortex structures (**Figure 6**). Combining both portions to an oblique field enforces both effects. One finds either larger shifts as well as stronger modulated flow structures (**Figure 6(e,f)**).

3.4 Resume

An external applied magnetic field (independent of its orientation) *stabilizes* the CCF (2-CCF) basic state: for a given value of the outer Reynolds number Re_2, the bifurcation thresholds $Re_{1,c}$ for vortex structures are shifted to higher values of the inner Reynolds number Re_1. This holds likewise for toroidally closed (w)TVF, as well for helical (w)SPI. Thereby, these shifts are typically linear in the squared field parameters s_x^2 or s_z^2, respectively. Moreover, stabilization is in general stronger for helical solutions (w)SPI than for toroidally closed ones (w)TVF. Additionally, an external applied magnetic field can, depending on its orientation, change the spatio-temporal structure of the classical TVF and of SPI. The reason is the stimulation of additional modes entering the axial and azimuthal Fourier decomposition of the flow. A pure axial magnetic field keeps the symmetry properties unchanged and does not change the flow structures. It only alter the flow amplitudes, i.e., the magnitude of up-shift in the onsets and magnitude of modes after bifurcation. On the other hand, as soon as a *finite transversal* field component is present (pure transversal or oblique field), all flow structures become always *3-dimensional* and therefore qualitatively different to the classical TVF and SPI states. In particular, they are wavy-like modulated and the pure structures do *not* exist anymore. Such a field component breaks the system symmetry and therefore alters either basic state (CCF changes to 2-CCF) and supercritical instabilities. As a result, one finds new solutions, field generated, *nonrotating* wTVF, contrary to the classical rotating ones. The vortex tubes of these wTVF are periodically expanded and constricted in azimuthal θ-direction, and this deformation pattern is *stationary*. In general, the bifurcation thresholds $\mathrm{Re}_{1,c}$ for non-axisymmetric, helical $m = 1$ (w)SPI are stronger modified in presence of a magnetic field than those for toroidally closed $m = 0$ (w)TVF. The upward shift towards larger values of Re_1 of the former is bigger than for the latter. One direct effect is the shift of the co-dimension point toward stronger counter-rotating frequencies. This effect has been found to show in axial, transversal, and oblique magnetic fields [6, 9].

Finally it is worth to mention that the magnetic field (transversal or oblique field) generated wTVF differ crucially from the classic wTVF [20, 33, 35, 45] that bifurcates *secondary* out of TVF at relatively high $\mathrm{Re}_{1,c}$ in absence of a magnetic field. The present generated wTVF arises via a *primary* bifurcation directly out of the 2-CCF basic state.

4. Agglomeration and elongational flow effects

4.1 Introduction and motivation

Aside the general difficulties to incorporate the magnetization of any magnetic field into the governing equations of motion, it is also crucial to consider the specific properties of the magnetic particles. However, most mathematical models describing the flow of complex magnetic fluids typically assume noninteracting magnetic particles with a *small* volume or only *point-size* objects. However, real ferrofluids consist of a suspension of particles with a *finite size* in an almost ellipsoid shape. Moreover, they involve *particle-particle interactions* resulting in agglomeration; in particular, they tend to form chains of various lengths. One possibility to include such effects and to come close to the realistic situation for ferrofluids is to consider the effect of *elongational flow*. Mathematically, such flow is incorporated by the symmetric part of the velocity gradient field tensor, which could be scaled by a so-called, material dependent, *transport coefficient* λ_2 [10, 37, 38]. As discussed in the previous section, magnetic fields tend to change the bifurcation threshold of the supercritical unstable flows and, depending on their orientation, resulting in structural modifications. Consider elongational flow, these modifications are expected to change the obtained states at least qualitatively.

4.2 Modified internal magnetic field

In order to consider such *elongational flow* (i.e. agglomeration, chain formation, etc.), one has to determine the relationship between the magnetization $\vec{M}$, the magnetic field $\vec{H}$, and the velocity $\vec{u}$. One possibility is to consider an additional dependence of the magnetization on the symmetric part of the velocity gradient $S = \left(\partial_i u_j + \partial_j u_i\right)/2$ [10, 38]. Thus, the magnetization (4) becomes modified to

$$\vec{M} - \vec{M}_{eq} = c_N \vec{G} \tag{11}$$

where $G = \vec{\Omega} \times \vec{H} + \lambda_2 S \vec{H}$.

Onsager symmetry relations then require [12, 39] that additional forces involving λ_2 appear in the off-equilibrium momentum balance of the ferrofluid.

Now, using the latest modified magnetization (9), the magnetization part in (1) can be eliminated, similar to previous discussion in Section 2.2. We thus have the following ferrohydrodynamic equation of motion (including elongational flow effects) [12]:

$$\left(\partial_t + \vec{u} \cdot \nabla\right) \vec{u} + \nabla p_M - \nabla^2 \vec{u} = -c_N^2 \left[\vec{H} \nabla \cdot \vec{G} + \vec{H} \times \nabla \times \vec{F}\right]/2 \qquad (12)$$

where p_M is the dynamic pressure incorporating all magnetic terms which can be written as gradients, i.e., $\nabla\left(\vec{H} \cdot \vec{H}\right)$ and $\nabla\left[\left(\vec{H}_{eq} - \vec{H}\right) \cdot \vec{M}\right]$. See also (5) for comparison *without* considered elongational flow.

Assuming that the internal magnetic field is equal to the external imposed magnetic field (it is known as a leading-order approximation [24], see also following section 5) which is sufficiently good for a 'first' numerical investigation for the effect of elongational flow. Therefore, (10) can be simplified:

$$\begin{aligned} \left(\partial_t + \vec{u} \cdot \nabla\right) \vec{u} + p_M - \nabla^2 \vec{u} = s_N^2 \Big\{ \nabla^2 \vec{u} - 2\lambda_2 \nabla \cdot S \vec{H} - H \\ \times \left[2\nabla \times \left(\vec{\Omega} \times \vec{H}\right) - \vec{H} \times \nabla^2 \vec{u} + \lambda_2 \nabla \times S\vec{H}\right] \Big\} \end{aligned} \qquad (13)$$

In this approach, the magnetic field and all the magnetic properties of the ferrofluid could be influenced by the velocity field, the transport coefficient λ_2, and the magnetic field Niklas parameters s_x and s_z as described before (7). However, it is worth to mention that in this approach, the material dependent transport coefficient λ_2 is a parameter [12], which can be scaled by the term describing the elongational flow and is *independent* of $\vec{H}$. Thus, it only appears in combination with the symmetric component of the velocity field tensor. However, real ferrofluids show $\vec{H}$ dependence in the microscopic aspect of the dynamics of ferrofluids with chains or nonspherical particles [25].

4.3 Linear stability behavior and analysis

Before we come to the influence of λ_2 onto the fully nonlinear Eqs. (11), it is worth to have a look at the linear marginal stability threshold of the different flow structures. Therefore, in the following, we first investigate the linear growth behavior of either stationary axisymmetric $m = 0$ (TVF) and nonaxisymmetric $m = 1$ (SPI) oscillatory vortex flow of a ferrofluid in the TCS that is placed in homogeneous axial magnetic fields $\vec{H}_{ext} = H_z \vec{e}_z$ (**Figure 1**). In the literature [4, 5, 7, 40–42], one can find many theoretical works that analyze (via linear stability analysis) the influence of rotational symmetric magnetic fields, i.e., axial, azimuthal, and radial ones on the flow of a ferrofluid in the TCS. However, for a symmetry breaking (e.g., transverse), magnetic field such linear stability analysis is *not known* for the corresponding basic state at such a field configuration. For a detailed description regarding the linearization around the CCF and solution of the linearized NSE, we refer to the Appendix in [43].

4.3.1 Marginal stability thresholds

In order to discuss how the transport coefficient λ_2 influences the growth of $m = 0$ and $m = 1$ vortex perturbations, we determine the stability of the CCF basic state against these perturbations, considering a magnetic field $\vec{H}_{ext} = H_z \vec{e}_z$, applied in pure axial direction.

Figure 9 presents the bifurcation thresholds in the $Re_2 - Re_1$ diagram for either axisymmetric, $m = 0$, and nonaxisymmetric, $m = 1$, vortex flow in magnetic fields at (a,b) $H_{ext} = 67.7\ kA/m$ and (c) $H_{ext} = 135.4\ kA/m$ for two different axial

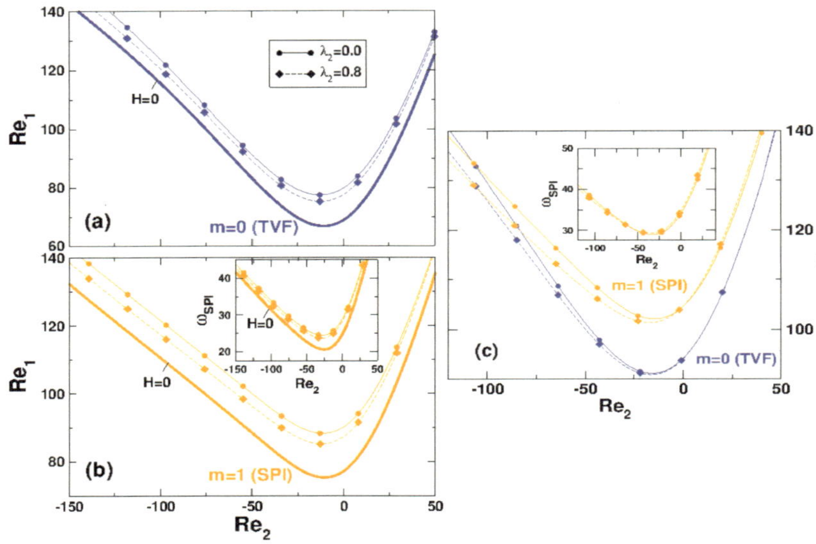

Figure 9.
Stability thresholds of the CCF basic state against vortex growth of (a) $m = 0$ *(TVF) and (b)* $m = 1$ *(SPI) in the* $Re_2 - Re_1$ *diagram of control parameters. Thin solid and dashed lines with symbols show the results for* $\lambda_2 = 0$ *and* $\lambda_2 = 0.8$ *for a magnetic field* $H_{ext} = 67.7kA/m$*. The axial wavenumber is* $k = 2.8274$*. (c) Same curves, with wavenumber* $k = 3.4558$ *and magnetic field is* $H_{ext} = 135.4kA/m$*. The lowest thick curve in (a) and (b) refers in each figure to the threshold in absence of a magnetic field* $H_{ext} = 0$*. Insets show corresponding SPI frequencies at the bifurcation thresholds.*

wavenumbers. Note that the representative axial wavenumber chosen in (a) $k = 2.8274$ is as detected in experiments [10]; (b) $k = 3.4558$. Insets in **Figure 9(b,c)** show corresponding SPI frequencies at the bifurcation thresholds. As already discussed in Section 3.3.4, the CCF basic state is stabilized against the two types of vortex perturbations in any magnetic field (the lowest thick solid curve in each plot in **Figure 9** refers to the case without any magnetic field H = 0 (c.f. **Figure 8(a)** for non-linear calculation)). Consider elongational flow ($\lambda_2 \neq 0$), the basic state becomes destabilized in comparison with the case $\lambda_2 = 0$. Regarding **Figure 9(a,b)**, both bifurcation thresholds for $m = 0$ and $m = 1$ vortices are shifted to *lower* values of Re_1. Thus, for the axial wavenumber of $k = 2.8274$, the λ_2 term *reduces* the general stabilization effect caused by an applied magnetic field. For parameters as presented in **Figure 9(a,b)**, this holds for the whole range of presented Re_2, even when the changes of the stability boundaries are quite weak, in particular for corotating cylinders. It is important to mention that it never over-compensate the field induced stabilization. **Figure 9(c)** illustrates in general a strong parameter dependence of the elongational flow effect. While at $Re_2 < 22$ for TVF and $Re_2 < 4$ for SPI, increasing λ_2 also has destabilizing effect, and for corresponding larger Re_2, the effect is just *opposite*. Increasing λ_2 even enforce the stabilization due to the applied magnetic field.

It is worth to mention that, in contrast to the stabilization of the CCF state by a magnetic field, the λ_2-induced modifications of the linear growth of vortices depend significantly on their axial wavenumber k as will be discussed below.

4.3.2 Magnetic field dependence of thresholds

Figure 10(I) illustrates the variation of the bifurcation thresholds $m = 0$ (TVF) and for $m = 1$ (SPI) with the magnetic field $\vec{H}_{ext} = H_z \vec{e}_z$, incorporating the elongational flow effect. The top row of figures refers to $Re_2 = 0$, and the bottom one to $Re_2 = -100$ and wavenumbers are $k = 2.8274$ (left) and $k = 3.1415$ (right), respectively. Ignoring the effect at small H_{ext}, all thresholds increase linearly with H_{ext}. The slopes are given by $\partial \mathrm{Re}_{1,stab}(H)/\partial H$ and vary slightly either with λ_2 and k. For outer cylinder at rest, $\mathrm{Re}_2 = 0$, (**Figure 10(I)(a)**) the slopes of the $m = 0$ and $m = 1$ curves are comparable, they differ significantly for counter-rotating cylinders, $\mathrm{Re}_2 = -100$ (**Figure 10(I)(b)**). For the latter, this results in intersections of the $m = 0$ and $m = 1$ stability boundaries. Such intersections have been observed also in experiments and also in full nonlinear simulations [6] with $\lambda_2 = 0$.

In general, modifications of the H_{ext} field dependence of the bifurcation thresholds due to λ_2 tend to increase, when the cylinders are stronger counter-rotating (**Figure 10(I)**). Re_2 becomes more negative, and thus, the strain increases. On the other side, they also depend on the axial wavenumber k (see also Section 4.3.4). Considering fixed k, the modifications due to λ_2 increase with increasing magnetic field H_{ext} while remaining qualitatively the same. However, the general tendency visible in **Figure 11**, considering elongational effect, is to shift the stability boundaries downward in Re_1. Concrete amounts depend on the parameters k and Re_2. Note that this relative small destabilizing elongational effect is always overcompensated by the significant stronger general stabilization effect of the magnetic field itself.

4.3.3 Frequencies of helical SPI states

Instead of looking at the growth rates—the real part of the eigenvalue of these solutions (previous Sections), one can also consider the marginal spiral frequency ω—the imaginary part of the marginal eigenvalue at the bifurcation threshold of $m = 1$ (SPI) vortex flow. **Figure 10(II)** shows the frequency, corresponding to the $m = 1$ bifurcation thresholds presented in **Figure 10(I)**. Main effect is the increase of the SPI frequencies with the field H_{ext}, accordingly to the increase of onset Reynolds numbers with H_{ext} (**Figure 10(I)**). Thus, the threshold frequencies are larger for $\mathrm{Re}_2 = -100$ than for $\mathrm{Re}_2 = 0$. In general, the effect induced by elongational flow is minor, significant smaller, and only secondary. Analog to the slightly down shift in the onsets (**Figure 10(I)**) also the SPI frequencies are slightly reduced (**Figure 10(II)**). Strong counter-rotating cylinders reduce in general the frequency for finite λ_2 (**Figure 10(II)**), while for larger wavenumbers, a finite elongational flow $\lambda_2 \neq 0$ at $\mathrm{Re}_2 = 0$ it can be slightly decreased.

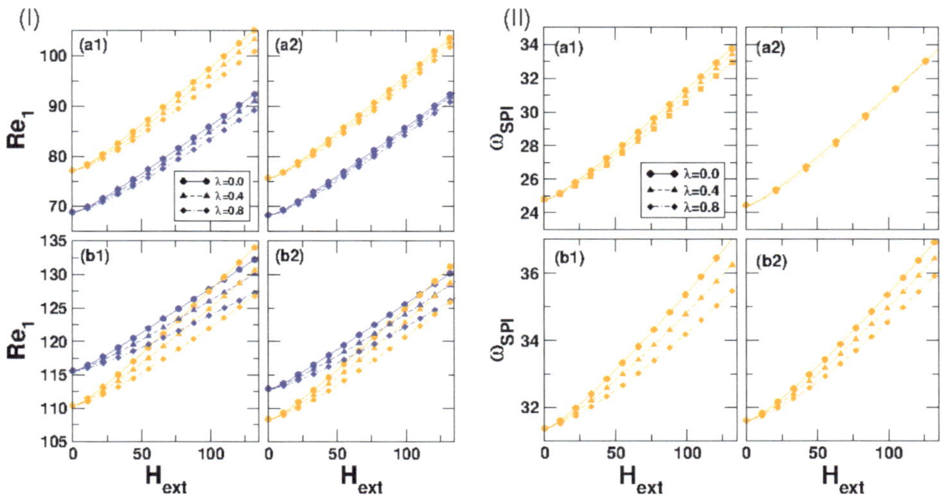

Figure 10.
(I) Variation of the (a) $m = 0$ (TVF) and (b) $m = 1$ (SPI) bifurcation thresholds and (II) corresponding SPI frequencies ω with magnetic field H_{ext} for different λ_2 as indicated. The top row refers to $\mathrm{Re}_2 = 0$ and the bottom one to $\mathrm{Re}_2 = -100$. The wavenumbers are (1) $k = 2.8274$ and (2) $k = 3.1415$, respectively.

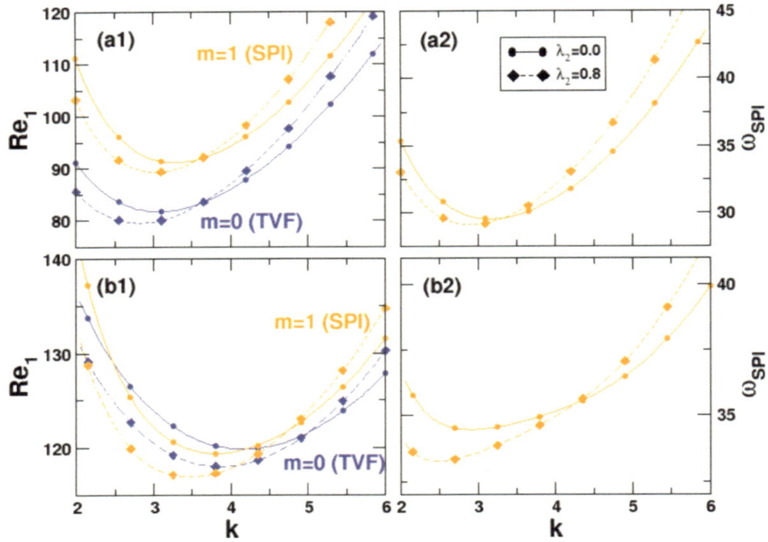

Figure 11.
(1) Stability boundaries of the CCF basic state against the growth of $m = 0$ *(TVF) (blue) and* $m = 1$ *(SPI) vortex flow (orange) in the* $k - \mathrm{Re}_1$ *diagram for* $\lambda_2 = 0$ *(solid lines) and* $\lambda_2 = 0.8$ *dashed lines. (2) corresponding marginal spiral frequencies* ω *at the* $m = 1$ *thresholds. Further parameters: (a)* $\mathrm{Re}_2 = 0$ *and (b)* $\mathrm{Re}_2 = -100$. *The (axial) magnetic field is constant with* $H_{ext} = 80kA/m$. *Results from linearized equations [43].*

A general observation is that all modifications of the spiral frequencies due to finite $\lambda_2 \neq 0$ are most pronounced for the smaller wavenumbers (here $k = 2.8274$ in **Figure 10**) in counter-rotating cylinders. The field dependence is virtually not much modified due to variations in λ_2, and the growth rate of the imaginary part of the marginal eigenvalue with increasing λ_2 is quite similar to that of its real part.

4.3.4 Axial wavenumber dependence

Following, we will first discuss the situation for a (moderately) larger wavenumber, taking $k = 3.4558$ as one specific, representative example. For larger value k (**Figure 9(c)**), the situation becomes more complicate. Incorporating the elongational flow scaled by the parameter λ_2 increases and decreases the growth rate, with respect to $\lambda_2 = 0$, and thus shifts the stability boundaries, up or down, depending on Re_2.

Figure 11 illustrates the relation between the wavenumber dependence of the stability boundaries and of the spiral frequencies. This is of particular interest in experiments, since the wavenumbers of some vortex structures show a strong variation when changing the magnetic field [9, 44]. Reindl *et al.* [9] also observed hysteresis between vortex flows with different k when increasing and decreasing the Reynolds numbers Re_1. The bifurcation thresholds (**Figure 11**, top row $\mathrm{Re}_2 = 0$ and the bottom row $\mathrm{Re}_2 = -100$; the (axial) magnetic field is fixed at $H_{ext} = 80\ kA/m$) for $m = 0$ and $m = 1$ vortex flow structures in the $k - \mathrm{Re}_1$ diagram change when the elongational term ($\lambda_2 \neq 0$) is considered in the linearized ferrofluid equations [43]. Right column of **Figure 11(2)** presents the wavenumber dependence of the corresponding marginal spiral frequencies at the onsets for $m = 1$ SPI.

Typically, the thresholds for $\lambda_2 \neq 0$ (in **Figure 11** the case of $\lambda_2 = 0$, lines with squares) are shifted downward in Re_1 at smaller wavenumbers k and upward at larger k. Resulting in a fixed magnetic field, the elongational effect destabilizes (stabilizes) the CCF state against growth of vortices with small (large) wavenumbers, either for $m = 0$ or $m = 1$ perturbations. Consequently, the stability onsets cross for different values of λ_2 ($\lambda_2 = 0$, $\lambda_2 = 0.8$ in **Figure 11**) at a specific value of k_{cross} the wavenumber.

It is worth to mention that the crossover values of k_{cross} for $m = 0$ and $m = 1$ perturbations also depend sensitive on other the control parameters, e.g., Re_2. Based on the observations in **Figure 11**, it is fair to assume that, depending on the wavenumber k, the γ-point (Section 3.3.4) will change.

Similar to the stability thresholds, also the spiral frequencies (**Figure 11(2)**) increase with increasing λ_2 when k is large and decrease for smaller k as a result of switching on the elongational effect. Thereby, the crossing wavenumber, k_{cross}, separating the increase and decrease in ω is slightly smaller than the one for crossing the marginal thresholds (**Figure 11(a1,b1)**). Following the speculation by Odenbach and Müller [10] on the physical origin of the elongational effect, one can give a reasonable explanation for such an observed wavenumber dependence of the λ_2 effect on the growth rates of vortex flow. In [10], the authors argue that the microscopic origin of a finite λ_2 might be attributed to the finite asphericity of the colloids, i.e., the presence of (short) particle chains in the ferrofluid. Small axial wavenumber k will allow these chains *more easily* to remain aligned in the direction of the magnetic field in vortex flows instead in a flow with large axial wavenumber and large azimuthal vorticity. Larger wavenumber means smaller wavelength and therefore more higher probability to *unalign* in the given field. Thus, since the magneto-rotational dissipation in the latter flow is larger than in the former one, it requires larger centrifugal forces, i.e., larger values of Re_1, to drive the growth of a large k flow. However, it is important to state that this argumentation also holds some weakness: In the presence of chains, one could expect also for small k vortices an increase of the threshold Reynolds number Re_1. On the contrary, incorporating the elongational term ($\lambda_2 \neq 0$) in the magnetization balance changes the momentum balance of the vortex flow such that the growth of vortices is enhanced for $k < k_{cross}$ (left in **Figure 10**) and reduced for $k > k_{cross}$ (right in **Figure 10**) relative to the reference case of $\lambda_2 = 0$.

4.4 Nonlinear analysis

4.4.1 Bifurcation thresholds

Following, we will consider the full nonlinear ferrohydrodynamical equation of motions incorporating elongational flow effects (11) in order to investigate the influence of a finite transport coefficient λ_2 at different applied magnetic fields.

The general modification in the bifurcation threshold of primary instabilities ((w)TVF) and ((w)SPI) due to finite values $\lambda_2 \neq 0$ remains qualitative as detected and described before in the linear stability analysis in pure axial field. However, the absolute values are slightly different. A direct comparison for axial field (**Figures 9-11**) results in variation about one to two percent between the calculated onsets using either the linear analysis or full nonlinear equations [6]. While the different field orientations result in different magnitudes of stabilization of the basic states CCF and 2-CCF, respectively, as discussed in Section 3.3.4, a finite value λ_2 does *not* depend on different field orientations. The variation effect with λ_2 is almost identical for all considered field configuration. However, in general, the modifications are slightly larger under oblique magnetic fields. Depending on other system parameters, e.g., wavenumber k, cylinder rotation speeds Re_1, Re_2, the effect is either stabilizing or destabilizing with respect to the situation $\lambda_2 = 0$ (**Figure 10(c)** and **Figure 11**). Considering the effect of elongational flow, the critical bifurcating point of primary centrifugal instability will be shifted either up or down. As detected in the linear analysis, the potential destabilizing effect of λ_2 *never overcomes* the general stabilizing effect of the external applied magnetic field, i.e., comparing $\lambda_2 = 0$ and any value $\lambda_2 \neq 0$.

4.4.2 Flow structure modifications

Aside the modification of the primary stability thresholds, there are further flow structural changes/modifications which can be observed for finite value λ_2. **Figure 11** shows, for different field configurations T: $(s_x = 0.1, s_z = 0)$, A: $(s_x = 0, s_z = 0.1)$, O: $(s_x = 0.1, s_z = 0.1)$, how the maxima of vorticity η vary for either the full flow structure $\Delta\eta{:=}\eta(\lambda_2 \neq 0) - \eta(\lambda_2 = 0)$ (**Figure 12(a)**), and reduced 2-fold subspace (**Figure 12(b)**) varies with λ_2 for the basic state configuration $(\mathrm{Re}_1 = 60, \mathrm{Re}_2 = 0)$. Remember, that only for $s_x = 0$, one has the classical CCF, and for any $s_x \neq 0$, a modified basic state is present [6, 9, 37] (i.e., symmetry breaking effect of transversal field). Although for a given configuration (A, T, O), the absolute maxima of these values are significantly different, and their relative differences Δ scale is *linear* with λ_2 starting at zero. Lines in **Figure 12** are calculated by linear regression fitting to the numerical data points. The effect of λ_2 is *decoupled* from the magnetic field configuration and with this the considered flow structure.

Either for classical "pure" CCF or field modified 2-CCF, the variation in $\max(\Delta\eta)$ grows linear with λ_2 (**Figure 12(a)**) (note the significant smaller amplitude in the $m = 2$ subspace). Further confirmation for this is the fact that an analog linear growth can be also found when considering other parameters, at which a supercritical flow state exists [37]. This holds for toroidally closed structures (w)TVF, as well as for helical flow states (w)SPI. Only the absolute values in $\max(\Delta\eta)$ are significant larger.

It is important to mention that the slopes in **Figure 12** (for $O(s_x = 0.1, s_z = 0.1)$) not only depend on the applied field orientation (O) but also depend on the *strength* of the applied magnetic field. With increasing the magnetic field strength s_x, s_z, the slopes become steeper. This scenario is illustrated, exemplary for 2-CCF basic state in transversal field configuration in **Figure 13**.

Moreover, similar characteristics hold for the different primary bifurcating structures. **Figure 14** elucidates this effect for the three different field configurations A, T, and O as indicated (see also **Figures 4** and **6** in [37] for another example considering a transversal magnetic field in finite system configuration). For a given field configuration, the slopes increase with λ_2, when increasing the magnetic field strength s_x. In general, slopes are smallest for transversal and largest for oblique fields, respectively. The variation in the slopes is continuous with increasing the field strength, but not linear; for larger field strength, the variation in the slopes becomes smaller.

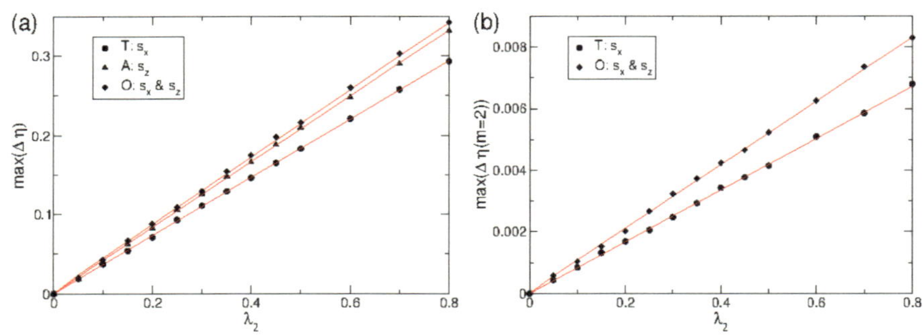

Figure 12.
Variation of vorticity maxima with λ_2 for (a) $\max(\eta) = \eta(\lambda_2) - \eta(\lambda_2 = 0)$ at full solution and (b) $\max(\eta(m = 2)) = \eta(\lambda_2) - \eta(\lambda_2 = 0)$ at only 2-fold symmetric subspace (empty for A) for basic flow (CCF and 2-CCF, respectively) at $\mathrm{Re}_1 = 60$, $\mathrm{Re}_2 = 0$ and different field orientations: Axial (A) $s_x = 0.0, s_z = 0.6$; transverse (T) $s_x = 0.6, s_z = 0.0$ and oblique (O) $s_x = 0.6, s_z = 0.6$. Lines are linear fits.

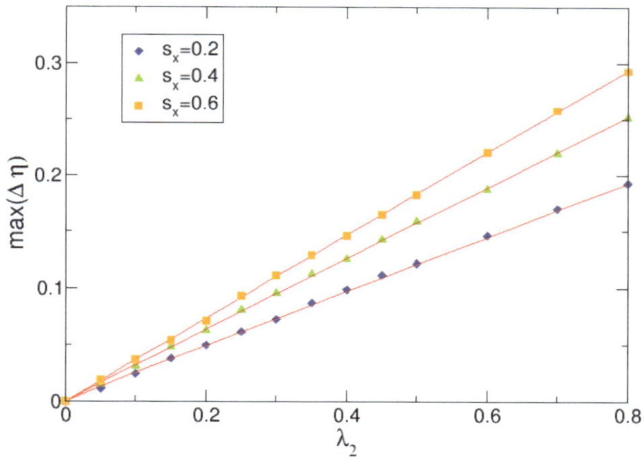

Figure 13.
*As **Figure 11** for basic state 2-CCF at* $\mathrm{Re}_1 = 60$, $\mathrm{Re}_2 = 0$. *Variation with* λ_2 *of vorticity maxima* $\max(\eta) = \eta(\lambda_2) - \eta(\lambda_2 = 0)$ *for transversal field (T) at different field strength* s_x *as indicated.*

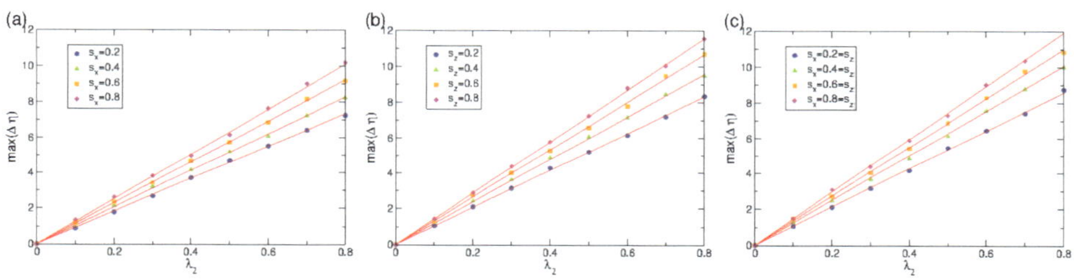

Figure 14.
Supercritical (w)TVF solution at $\mathrm{Re}_1 = 100$, $\mathrm{Re}_2 = 0$. *Variation of vorticity maxima* $\max(\eta) = \eta(\lambda_2) - \eta(\lambda_2 = 0)$ *for (a) transversal (T), (b) axial (A) and (c) oblique (O) magnetic field (strength as indicated). Lines are linear fits.*

4.4.3 Wavenumber modification and selection

As seen in the former discussion (section 4.3.2), an elongational flow can have strong effect on the axial wavenumber k. Consider the full equations (and periodic boundary conditions), one obtains similar results as presented in **Figure 10** for linear analysis of the marginal stability thresholds for TVF and SPI. As described before, the agreement between the results of linear and full equations is very close (they differ less than 2 percent).

However, in order to see such wavenumber modification from another perspective, it makes sense to study a finite system with sufficient large aspect ratio enclosed by fixed end plates (assumed to be stationary) on top and bottom. As in a finite system, the axial wavenumber k is *natural selected* by the system itself (depending on the various system parameters), it is possible to see the direct influence, considering a finite value λ_2 on to the flow structures and their axial wavenumbers k.

Starting with a classical TVF as initial flow state, with $n = 22$ vortices, which corresponds to an axial wavenumber $k = 3.41$ in the bulk (illustrated on the most left border in **Figure 15**), we consider a situation with an applied pure transversal magnetic field which is continuously increased for two values, either ignoring any elongational flow effect, $\lambda_2 = 0$ (**Figure 14(a)**), or considering such effect $\lambda_2 = 0.2$ (**Figure 15(b)**) (note 0.2 is quite small). In any case, either for $\lambda_2 = 0$ and for $\lambda_2 = 0.2$, by increasing the magnetic-field strength, s_x, the number of vortices is reduced, which leads to an increase in the axial wavelength (i.e. decrease in axial wavenumber k) due to *elimination* of vortices in the bulk. Therefore, the corresponding axial wavenumber k in the bulk becomes reduced. Remember that the initial state is a classical, stationary TVF, which becomes modified to stationary wTVF as soon as $s_x \neq 0$. For the field generated wTVF, the wavy-like modulation is not visible in **Figure 15**. However qualitative, the same scenario also holds for either axial and oblique magnetic fields, with changes in the corresponding parameters s_x, s_z at which the wavenumber k changes, i.e., vortices disappear.

In general, the exact field strength s_x for vortex elimination and/or the number of destroyed vortices depends on λ_2. In the example of **Figure 15**, the initial flow has 22 vortices with an axial wavenumber $k = 3.41$. Neglecting elongational effects $\lambda_2 = 0$ (**Figure 15(a)**) until the parameter s_x approximately reaches the value about 0.7, the flow remains stable, and the number of vortices in the flow can be reduced to $n = 20$ with $k = 3.04$. Further increasing s_x to about 0.85, the number of vortices can be more reduced to $n = 16$ with $k = 2.58$. Hereafter, no further changes in the number of vortices n appear, before reaching the boundary threshold, $s_x \approx 0.9$. Consider elongational flow $\lambda_2 = 0.2$ (**Figure 15(b)**), the behavior of the vortex state is similar to the case neglecting such effects. The flow remains stable before the parameter s_x reaches about 0.75. At the same time, the number of vortices is also reduced in analog manner to $n = 20$ with $k = 3.04$. Hereafter, as before, the number of vortices n remains unchanged until the parameter s_x increases to the boundary threshold, $s_x \approx 0.95$. Consider values $\lambda_2 > 0.7$, the flow structure (for given parameters) is not affected by the magnetic-field strength. A flow with $n = 22$ vortices corresponding to axial wavenumber $k = 3.41$ in the bulk is remaining in the bulk.

An interesting phenomenon, which also has been reported in experimental studies [9] as in numerical simulations, is the fact that one or two vortex pairs can be eliminated. However so far, there is no physical explanation for this observation/phenomenon. Analog to the presented scenario in **Figure 15** for pure transversal magnetic field, one also finds similar reduction scenarios for pure axial and oblique magnetic fields. As to expect, the absolute values of field strength, s_x, s_z, at which the elimination of vortices happen is different, depending on the applied field.

4.5 Resume

Real ferrofluids consist of a suspension of particles with finite size and mainly almost ellipsoid shape, as well as there are particle-particle interactions that tend to

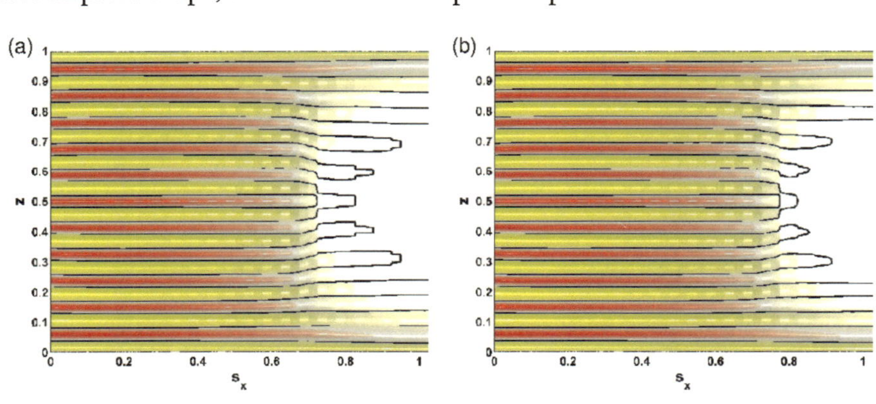

Figure 15.
Contours of the radial velocity component u at mid-gap with variation of field strength s_x at $Re_1 = 88.35$, $Re_2 = 0$ for (a) $\lambda_2 = 0$ and (b) $\lambda_2 = 0.2$. Left border in each plot gives the initial state, a classical TVF with $n = 22$ vortices corresponding to an axial wavenumber $k = 3.41$ in the bulk. Note that for $s_x = 0$, only wavy-like modified flow states exist. Red (yellow) [dark gray (light gray)] corresponds to positive (negative) values. The max (min) level is ± 9.98.

form chains of various lengths and tend to agglomerate in general. One possibility in order to make a more realistic approach for ferrofluids and to come close to such a configuration in a real ferrofluids is to consider the effect of elongational flow. Such a flow becomes incorporated by the symmetric part of the velocity gradient field tensor in the nondimensional ferrohydrodynamic equations of motion (11) and scales by a material dependent transport coefficient λ_2. By such an assumption, the latter is given by a pure scalar number, describing the strength/influence of all former mentioned properties.

Consider either linear stability analysis or fully nonlinear equations, one finds that incorporating such elongational flow can have either stabilization or destabilization effect (i.e., modification in the supercritical primary bifurcation thresholds) with respect to the reference situation in absence of such flow ($\lambda_2 = 0$). The modifications due to finite value λ_2 are qualitative similar and in the same range for different considered magnetic fields, slightly largest in case of an oblique field compared to pure axial or transversal ones. Typically, the modifications due to $\lambda_2 \neq 0$ are larger for helical, nonaxisymmetric, $m = 1$, (w)SPIs than for toroidally closed, axisymmetric, $m = 0$, (w)TVF. Thus, the regim of primary stable bifurcating (w)TVF increases. This observation is similar and goes in hand with the previously described general stabilization of any applied magnetic field (Section 3). There we saw that the stability thresholds of helical flow states are shifted further than those of toroidal closed flow structures. However, it is important and fundamental to state that the elongational effect $\lambda_2 \neq 0$ is *never* big enough to *over compensate* the general, field-induced stabilization. Only the magnitude/amplitude in shift can be reduced. Further, it is worth to mention that linear stability analysis is limited to axial magnetic fields; as for fields with finite transversal component, no basic state can be derived, simulations of full nonlinear equations confirm similar effects. The modifications due to λ_2 are sensitive to various different system parameters (e.g., Reynolds numbers, radius ratio, etc.), in particular sensitive to the axial wavenumber k. Here, smaller k are strongly effected than larger ones. Thus, considering finite size TCS, a considered elongational flow results in modifications of flow structures in the sense of changing number of vortices in the bulk, which coincides with a different wavenumber selection. Thus, depending on the particular wavenumber, the presence of any elongational effect $\lambda_2 \neq 0$ can be either stabilizing or destabilizing for the basic state. In finite system, this results in different selection and/or modification of number of vortex pairs within the annulus, depending on the parameter.

Regarding the flow structural properties, a finite transport coefficient, $\lambda_2 \neq 0$, does *not* change the qualitative shape of the flow pattern. Only quantitative differences can be observed. Essentially, all structural properties remain as generated/induced due to the external applied magnetic field. Thus for instance, the azimuthal mode-2 symmetry originated in the symmetry-breaking transverse magnetic field is preserved; only the strength of different flow quantities is modified (in general they increase) according to finite λ_2. Mainly, the flow states become *linearly enforced*, e.g., speaking in quantity of the azimuthal vorticity, which holds for both the basic flow (CCF, 2-CCF) and the primary instabilities the supercritical centrifugal unstable flows ((w)TVF, (w)SPI).

5. Modified field dependence - internal magnetization

5.1 Introduction and motivation

Aside the properties and behavior of the magnetic particles itself and their interaction with the applied magnetic field (see last sections), the correct assumption of the applied magnetic field is another crucially point in describing magnetic flows and their dynamics. The frequently used, most common assumption is to consider the internal magnetic field within the magnetic fluid (e.g., a ferrofluid) to be *equal* to the external applied field. Following, we will see that this is only a leading-order approximation. By accounting for the ferrofluid's magnetic susceptibility, a uniform externally imposed magnetic field is modified by the presence of the magnetic fluid (ferrofluid) within the annulus. The modification to the magnetic field appears in an radial dependence $1/r^2$ and their magnitude scales with the susceptibility χ. To be concrete, such modification to the imposed magnetic field can be substantial, also for ferrofluids typically used in laboratory experiments. These have significant consequences on the structural properties and stability of the basic states, as well as on the supercritical (primary) bifurcating solutions.

5.2 Modified internal magnetic field

As already seen in the previous sections, for solving the equation of motion (1), we have to consider the magnetic field within the annular gap between the cylinders. In the last sections, we used the simplest approach with the assumption to take the magnetic field to be identical to the applied external field [4, 6]. However, this simple approach is only a leading-order approximation, and depending on the magnetic susceptibility of the ferrofluid, the magnetic field in the gap is modified.

Assuming infinitely long cylinders, the magnetic boundary conditions are

$$\vec{H} = \vec{H}_{ext} - M_r \vec{e}_r \tag{14}$$

at $r = r_1$ and $r = r_2$, where $\vec{H}_{ext}$ is the homogeneous external applied magnetic field in the absence of the ferrofluid-filled annulus and M_r is the radial component of the magnetization $\vec{M}$.

The following ansatz satisfies a solenoidal field, $\nabla \cdot \vec{H} = 0$:

$$\begin{aligned} \vec{H} &= \vec{H}_{ext} + \left[\left(a_1 - b_1/r^2\right)\cos(\theta) + \left(a_2 - b_2/r^2\right)\sin(\theta)\right]\vec{e}_r \\ &\quad + \left[\left(a_2 + b_2/r^2\right)\cos(\theta) - \left(a_1 - b_1/r^2\right)\sin(\theta)\right]\vec{e_\theta} \end{aligned} \tag{15}$$

and then the boundary conditions defining the external field are.

$\vec{H}\,(r = r_1) = \vec{H}_{ext} + \left[\left(a_1 - b_1/r_1^2\right)\cos(\theta) + \left(a_2 - b_2/r_1^2\right)\sin(\theta)\right]\vec{e_r} + \left[\left(a_2 + b_2/r_1^2\right)\cos(\theta) - \left(a_1 - b_1/r_1^2\right)\sin(\theta)\right]\vec{e_\theta}$ and $\vec{H}\,(r = r_2) = \vec{H}_{ext} + \left[\left(a_1 - b_1/r_2^2\right)\cos(\theta) + \left(a_2 - b_2/r_2^2\right)\sin(\theta)\right]\vec{e_r} + \left[\left(a_2 + b_2/r_2^2\right)\cos(\theta) - \left(a_1 - b_1/r_2^2\right)\sin(\theta)\right]\vec{e_\theta}$.

To calculate the four unknown constants (a_1, a_2, b_1, b_2) from the boundary conditions (12), we need the radial component of the magnetization which results from substituting (13) into the magnetization (3):

$$\begin{aligned} M_r = &\left[\chi H_{ext}^T + \chi(a_1 - b_1/r^2) - c_N\Omega(a_2 + b_2/r^2)\right]\cos(\theta) \\ &+ \left[c_N\Omega H_{ext}^T + \chi(a_2 - b_2/r^2) + c_N\Omega(a_1 + b_1/r^2)\right]\sin(\theta), \end{aligned} \tag{16}$$

where H_{ext}^T stands for the transverse component of $\vec{H}_{ext}$. Using the continuity of the magnetic field condition on the cylinders, the coefficients can be calculated. Substituting them into (13), finally the resulting magnetic field is given by

$$\vec{H} = -\left(2\mathrm{H}_{ext}^{T}/K\right)\left(r_1/r^2\right)\left\{[\chi\cos(2\theta) + \Omega c_N \sin(2\theta)]\vec{e_x} + [\chi\sin(2\theta) + \Omega c_N \cos(2\theta)]\vec{e_y}\right\}, \tag{17}$$

where $K = (2+\chi)^2 - \chi^2\eta^2$. A special case of (15) when $\Omega c_N = 0$. This corresponds to an equilibrium magnetization and has been already observed in an earlier work by Odenbach and Müller [38]. They considered the limit of an equilibrium magnetization and used a similar derivation of the modification of the magnetic field for a ferrofluid between two cylinders.

Based on (15), one can use the following magnetic field in the equation of motion:

$$\vec{H} = 2\frac{(2+\zeta)}{K}H_x\left[\left(1-\zeta/r^2\right)\cos(\theta)\vec{e_r} - \zeta/r^2\sin(\theta)\vec{e_\theta}\right] + H_z\vec{e_z} \tag{18}$$

the dimensionless scalar parameter

$$\zeta = \frac{\chi}{2+\chi}r_1^2 \tag{19}$$

characterizes the strength of the radial field dependence ζ. The ferrohydrodynamical equation of motion (6) and the Niklas parameters s_x and s_z (7) remain unchanged/untouched from such considered radial field dependence.

5.3 Flow structural modifications

5.3.1 Primary instabilities

As we learned in previous sections, a magnetic field with a finite transverse component modifies either basic state or primary bifurcating flow structures in a wavy-like manner. More details of these states have been reported in numerical simulations [6], for which a homogeneous internal magnetic field equal to the external applied one has been supposed, as well as in real experimental studies [9]. The question now is how does the former derived radial field dependence ζ affects the stability of the basic states (CCF and 2-CCF) and the characteristics of the bifurcating solutions ((w)TVF and (w)SPI).

We begin our discussion by considering a purely axial applied magnetic field (A), with fixed field strength $s_z = 0.6$, which leaves the flows, in particular its structural properties unaffected (TVF and SPI). **Figure 16** shows the influence of the radial field parameter, ζ, (17) (exemplary chosen (a) $\zeta = 0$, (b) $\zeta = 0.4$ and (c) $\zeta = 0.8$), on the bifurcation thresholds for TVF and SPI in an axial magnetic field. As in absence of any field dependence, $\zeta = 0$ (**Figure 16(a)**), the bifurcations continue being supercritical. Main effect on the bifurcation curves is that with increasing ζ, the thresholds for the onset of both TVF and SPI are shifted to higher Re_1 (**Figure 16(b,c)**). Thereby, the threshold for the toroidally closed TVF being shifted significantly further (to larger values Re_1). Note the different scaling on the abscissa in **Figure 16(c)**. Aside from the enhanced stabilization of the CCF basic state on top of the field introduced stabilization itself, one observes an important consequence to the relative shifts in the two bifurcation curves. Consequently, the bifurcation curve of TVF is shifted more than the one of SPI, the co-dimension-two point, γ, is also shifted to more positive Re_2 (c.f. **Figure 15(a,b)**). Interestingly, this is just the *opposite* way of the shift with increasing field strength s_z in a pure axial magnetic field (the same holds for s_x in transversal field), which we saw in **Figure 7**

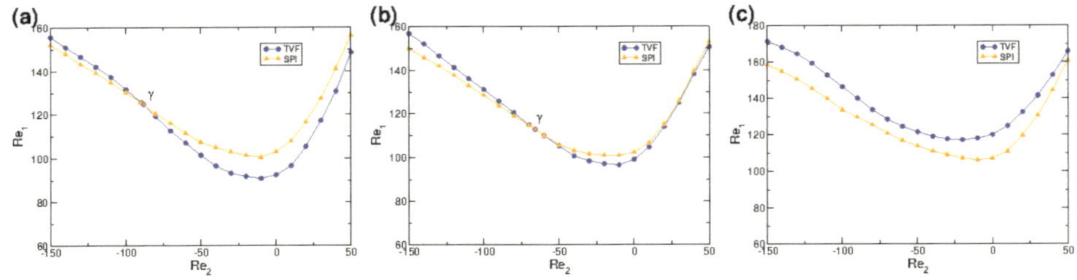

Figure 16.
Bifurcation curves for the onset of TVF (blue) and SPI (orange) in $\mathrm{Re}_2 - \mathrm{Re}_1$ *diagram for different* ζ*-values (a)* $\zeta = 0$*, (b)* $\zeta = 0.4$ *and (c)* $\zeta = 0.8$ *with an applied axial magnetic field (A) with* $s_z = 0.6$*. ((a) is a replot of* ***Figure 7(c)*** *for comparison). Note the different scale on the ordinate in (c). Simulation results of full nonlinear equations (11) consider a short periodic TCS with axial wavenumber* $k = 3.927$ *(same hold for* ***Figures 15*** *and* ***16****).*

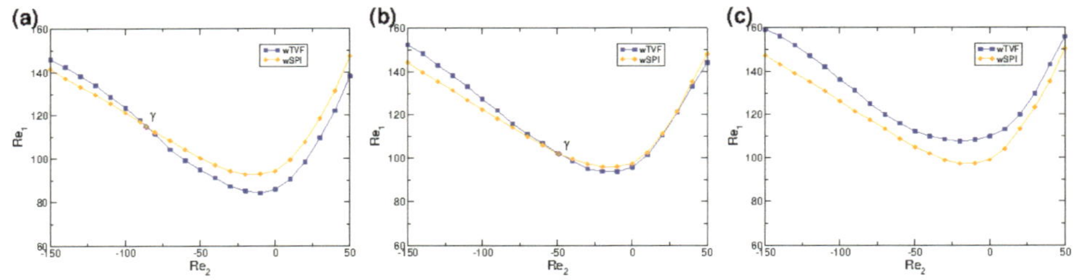

Figure 17.
Bifurcation curves for the onset of wTVF (blue) and wSPI (orange) in parameter space for different ζ *values (a)* $\zeta = 0$*, (b)* $\zeta = 0.4$*, and (c)* $\zeta = 0.8$ *with an applied transversal magnetic field (T)* $s_x = 0.6$*. ((a) is a replot of* ***Figure 7(b)*** *for comparison).*

with moving the γ-point to the left toward more negative values in Re_2. Consider larger values, e.g., $\zeta = 0.8$ (**Figure 15(c)**), one sees that the co-dimension-two γ-point is even shifted to $\mathrm{Re}_2 > 50$ out of the considered parameter space. As a result, the primary instability is always the helical SPI, even with the two cylinders in strong co-rotation. This observation is crucially different from classical TCS, both without magnetic fields and also with assumption of an applied uniform magnetic field. Thus, the radial field dependence, induced by the susceptibility of the ferrofluid-filled annulus, is seen to cause major changes in the quantitative and qualitative characteristics of the flow instabilities.

Consider a transversal magnetic field, field strength $s_x = 0.6$ (**Figure 17**) instead of an axial one, the variation by increasing ζ remains virtually and qualitatively the same, aside that now all structures are wavy-like modulated (2-CCF, wTVF, wSPI). One finds the same preference, in the sense of stronger upshifts for toroidally closed wTVF with the same result of a shift in the γ-point to the right to larger values in Re_2. Thus also here, for sufficiently strong ζ, e.g., $\zeta = 0.8$ in **Figure 17(c)**, the helical wSPI is the primary stable bifurcating solution in all presented parameter space. Comparing both, pure axial and pure transversal field (at same field strength), the up-shift, i.e., the stabilization is slightly larger in axial fields. This holds for toroidally closed (w)TVF and helical (w)SPI solutions.

Finally, this effect is even more pronounced when an oblique magnetic field is applied as illustrated in **Figure 18** for field strength $s_x = 0.6, s_z = 0.6$. Note also the different scale on the ordinate in **Figure 17(c)** highlighting the stronger shifts due to such field configuration. One can speculate that this enforcement is due to the axial component of the oblique field leading to a base state with w having axial shear, and the radial field dependence ζ enhances the w component of the 2-CCF base state. The combination of the base state azimuthal shear and this axial shear

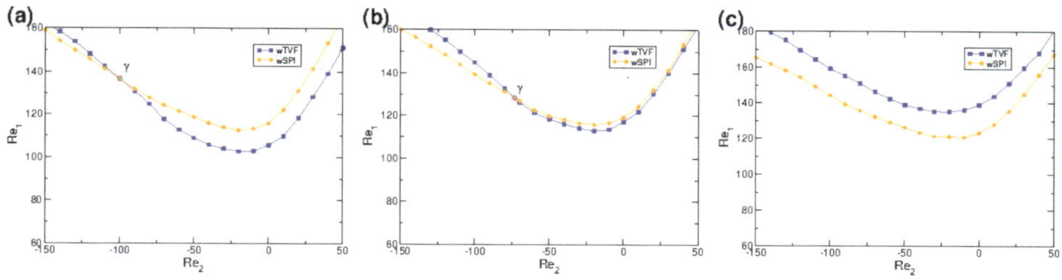

Figure 18.
Bifurcation curves for the onset of wTVF (blue) and wSPI (orange) in $\mathrm{Re}_2 - \mathrm{Re}_1$ *diagram for different* ζ*-values (a)* $\zeta = 0$*, (b)* $\zeta = 0.4$*, and (c)* $\zeta = 0.8$ *with an applied oblique magnetic field (O) with* $s_x = 0.6 = s - z$*. ((a) is a replot of* ***Figure 7(d)*** *for comparison). Note different scaling in (c).*

favors the helical instability. Close to the onset, the variation of the maxima of all flow components u, v, w is almost linear.

5.3.2 Modified flow structures

Figure 19 illustrates the flow structural properties of wTVF and wSPI in either a transverse magnetic field T (a,b) and an oblique magnetic field O (c,d), both with assumed radial field dependence $\zeta = 0.8$. Note, the flow structures of TVF and SPI under a pure axial applied magnetic field are not shown, as they virtually do not differ from the ones in absence of a magnetic field. Only the absolute values (e.g. in η) are different. Comparing these flow structures to the ones for $\zeta = 0$ cases as earlier presented in **Figure 6** (Section 3.3.3), one sees the modulations in all structures to be *enhanced/enforced*. In general, the waviness (axial variation in azimuthal direction) of

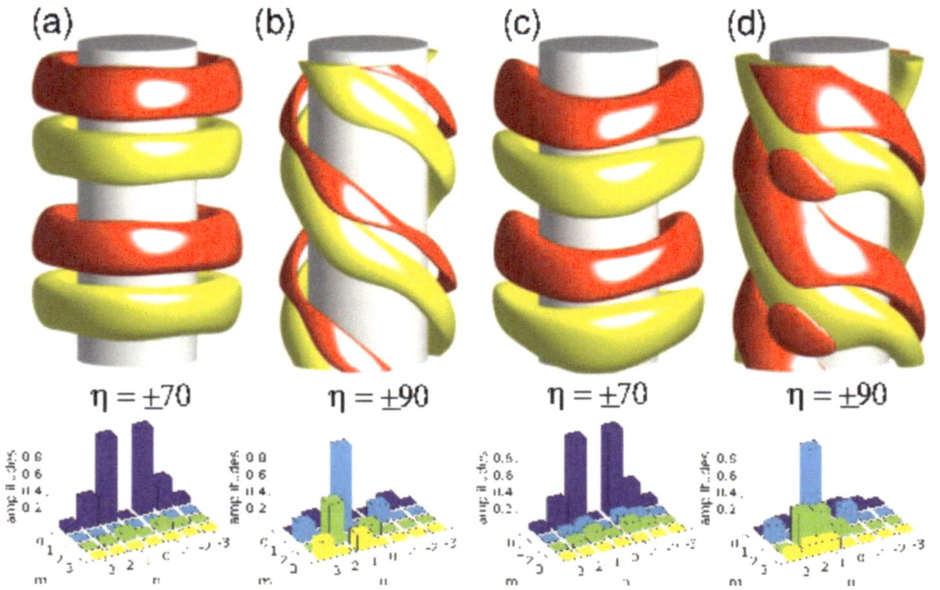

Figure 19.
(See also ***Figures 2*** *and* ***6*** *for comparison to flow states without radial field dependence. The radial field dependence is considered to be* $\zeta = 0.8$*. Top row: Azimuthal vorticity isosurfaces* η *over two axial wavelengths with pure transversal applied magnetic field (T): (a) wTVF at* $\mathrm{Re}_2 = 0$, $\mathrm{Re}_1 = -130$*, (b) wSPI at* $\mathrm{Re}_2 = -150$, $\mathrm{Re}_1 = 190$*; and with oblique applied magnetic field (O): (c) wTVF at* $\mathrm{Re}_2 = 0$, $\mathrm{Re}_1 = -150$*, and (f) wSPI at* $\mathrm{Re}_2 = -150$, $\mathrm{Re}_1 = -200$ *Red (yellow) isosurfaces correspond to positive (negative) values as indicated. The bottom row shows the mode amplitudes* $|u_{m,n}|$ *of the radial velocity field u corresponding to the structures above over the m-n-plane.*

all flow structures is enforced/enhanced by the radial dependent magnetic field. However, the field-induced wTVF always remains a *stationary*, *nonrotating*, and *phase-pinned* structure, irrespective of whether the radial field dependence is considered or not. On the other hand, wSPIs bifurcate as degenerated solutions, either as a left- or right-winding SPI, depending on initial conditions. Their corresponding frequencies remain mainly constant over a wide range and do not differ much (about 20% or less) from those of SPI. Comparing the stimulated modes for flow structures when $\zeta = 0.8$ (bottom in **Figure 19**) to those with $\zeta = 0$ (bottom in **Figure 6**), one sees general larger mode amplitudes in the additional stimulated modes for consideration of an internal field dependence ($\zeta \neq 0$). These larger amplitudes in the field induced modes are also the reason for the increase in waviness.

5.4 Resume

Applying a magnetic field across a container of ferrofluid changes the ferrofluid's susceptibility and results in modification of the magnetic field structure within the container. So far, most theories commonly neglect such modification, either for convenience or simplicity, and at the end, there is also possibility of numerical simulations in appropriate time or feasibility at all.

However, it is a matter of fact that an external applied magnetic field is modified, depending on the magnetic susceptibility. Experimental studies [10, 38] confirmed the modifications also for susceptibilities of commonly used ferrofluids. Consider such modifications to the imposed magnetic field effects the basic state, its stability, as well as the primary (supercritical) bifurcating solutions. This holds for any field configuration, with only differences in the strength of modifications, i.e., the stabilization (its magnitude; up-shift, move of the bifurcation thesholds to larger control parameters). On the other hand, the flow structural properties remain qualitative unaffected by such an radial field dependence, only change in quantitative manner. They are determined only by the general kind and orientation of the applied magnetic field. Thus, a pure axial field does not have any effect on the flow structural properties, leaving them as in absence of any field, i.e., pure TVF and pure SPI. As described before, the presence of a finite transversal field component results in a wavy-like modulated flow state. This is independent/untouched if the magnetic field is considered to be homogeneous or inhomogeneous with a radial field dependence. However, consider stronger modifications to the imposed magnetic field, i.e., larger values of the radial field dependence parameter ζ have the effect of further strengthening the already present waviness of the structures. This holds for (w)TVF and (w)SPI alike. An important observation is the different influence of the radial field dependence ($\zeta \neq 0$) onto topological closed and helical flow states. In general, the effect of modification in the primary bifurcation thresholds is larger for toroidally closed flow structures compared to helical ones. As a result, their intersection point, i.e., co-dimension γ-point is moved toward positive Re_2, which is just the opposite direction, as it moves with increasing the magnetic field strength (for any given field configuration, e.g. **Figure 7**). The parameter space, the area for which (w)SPI bifurcate primary stable out of the basic state, is growing, even into region of co-rotating cylinders, i.e. $Re_2 > 0$, making wSPI the favorable flow state in the system.

6. Conclusions

Although there has been significant progress in the understanding of ferrofluidic flows in the past decades, their fully theoretical understanding is still limited. A key reason for this is computational high-expensive simulations, which even increase with continuation in refining the models to make them more realistic/appropriate, e.g., finite size effects, internal field modifications, etc. However, these model improvements are essential and necessary to uncover the full potential of ferrofluids, which already are used in a wide field of applications. The use in the car industry within antishocks to provide adjustable damper settings is probably one of the most prominent examples. However, either experiments and numerics are essential to mutually prove each other and combined to provide new inside of potential applications.

This chapter deals with a rotating ferrofluid between two concentric cylinders, i.e., ferrofluid in Taylor-Couette geometry. Therefor we consider different improvements in the model describing such magnetic particles in a flow under external applied field. Concluding the discussions, the main points can be summarized as follows:

- Any applied magnetic field changes the bifurcation thresholds of primary solutions—the thresholds of both primary solutions, TVF and SPI, are shifted to larger parameter values—the *basic state becomes stabilized* against perturbations.

- Magnetic fields with a finite transversal component break the system symmetry and renders all flow states to be *inherently* 3D. As a result, the pure flow states TVF and SPI cease to exist and are substituted by their wave-like cousins, wTVF and wSPI. Thereby, the generated wTVF differs crucially from the classical ones, which rotate in azimuthal direction. Instead, the magnetic field induced wTVFs are *nonrotational, phase-pinned* flow solutions and develop *belly shape structure* at specific azimuthal positions.

- Assuming *elongational flow*, in order to consider *finite-size effect* of the magnetic particles and their interaction (agglomeration, particle-particle interaction and chain formation procedures), one detects further changes in the bifurcation threshold of the solutions. Also, this can result in further *stabilization* or *destabilization.* However, the effect of such elongational flow is *never* big enough to over-compensate the general, field-induced stabilization. Only the magnitude/amplitude varies. The flow states remain structural as in absence of elongational flow.

- Accounting for the changes in ferroluid's susceptibility in the container due to an external magnetic field, the real internal field can be approximated by $1/r^2$. As before the flow, structural properties remain qualitative similar. Bifurcation thresholds are shifted upward, toward larger Re_1, whereby toroidally closed (w)TVFs are in general preferred and shifted more than helical (w)SPI. The result is an increase in the regime of primary bifurcating (w)SPI.

A. Appendix

A.1 Numerical method

There are various different ways to solve the ferrohydrodynamic equations of motions (Eqs. (5) and (10)), the latter including elongational flow). All here presented results have been obtained by a code (G1D3 [6, 7]), which combines a finite difference method of second order in (r,z) and time (explicit) with spectral decomposition in θ [6]:

$$f(r,\theta,z,t) = \sum_{m=-m_{max}}^{m_{max}} f_m(r,z,t)e^{ink\theta}, \tag{20}$$

where f denotes one of $\{u, v, w, p\}$. For the parameter regimes investigated here, $m_{max} = 8$ provides adequate accuracy. For discretization, a forward time, centered space (FTCS) algorithm is used on staggered grids in the $r - z$ plane following the procedure of Hirt et *al.* [45]. In particular, a homogeneous grid with discretization length $\delta_r = \delta_z = 0.05$ and time steps $\delta t < 3800$ has been used. Here, δt defines the time step between two iterations via FTCS algorithm of the system of coupled equations for the amplitudes $f_m(r,z,t)$ of the azimuthal normal modes $-m_{max} \leq m \leq m_{max}$. For diagnostic purposes, the complex mode amplitudes $f_{m,n}(r,t)$ were calculated obtained from a second axial Fourier decomposition:

$$f_m(r,z,t) = \sum_n f_{m,n}(r,t)e^{inkz}, \tag{21}$$

where $k = 2\pi/\lambda$ is the axial wavenumber.

Acknowledgments/Funding

S. Altmeyer is a Serra Húnter Fellow. This work was supported by the Spanish Government grant FIS2017–85794-P.

Conflict of interest

The author declares no conflict of interest.

Nomenclature and abbreviations

TCS	Taylor-Coutte System
NSE	Navier–Stokes equation
CCF	Circular Couette flow [basic flow]
2-CCF	2-fold CCF [basic flow in transversal field]
TVF	Taylor vortex flow
SPI	Spiral vortex flow (spiral)
wTVF	wavy TVF
wSPI	wavy SPI

DOI: http://dx.doi.org/10.5772/intechopen.80301

Author details

Sebastian Altmeyer
Castelldefels School of Telecom and Aerospace Engineering, Universitat Politècnica de Catalunya, Barcelona, Spain

*Address all correspondence to: sebastian.andreas.altmeyer@upc.edu

IntechOpen

© 2018 The Author(s). Licensee IntechOpen. This chapter is distributed under the terms of the Creative Commons Attribution License (http://creativecommons.org/licenses/by/3.0), which permits unrestricted use, distribution, and reproduction in any medium, provided the original work is properly cited.

References

[1] Rosensweig RE. Ferrohydrodynamics. Cambridge University Press, Cambridge; 1985. 344 pp. Journal of Fluid Mechanics, 200, 597-599 DOI: 10.1017/S0022112089220773

[2] Hart JE. Dynamics of Atmospheres and Oceans. 2006;**41**. DOI: 10.1016/j.dynatmoce.2006.03.001

[3] Hart JE, Kittelman S. Dynamics of Atmospheres and Oceans. 2006;**41**:139. DOI: 10.1017/S0022112075001309

[4] Niklas M. Zeitschrift fuer Physik B: Condensed Matter. 1987;**68**:493. DOI: 10.1007/BF01471080

[5] Niklas M, Müller-Krumbhaar H, Lücke M. Journal of Magnetism and Magnetic Materials. 1989;**81**:29. DOI: 10.1016/0304-8853(89)90225-4

[6] Altmeyer S, Hoffmann C, Leschhorn A, Lücke M. Physical Review E. 2010;**82**:016321. DOI: 10.1103/PhysRevE.82.016321

[7] Leschhorn A, Lücke M, Hoffmann C, Altmeyer S. Physical Review E. 2009;**79**:036308. DOI: 10.1103/PhysRevE.79.036308

[8] Reindl M, Leschhorn A, Lücke M, Odenbach S. Journal of Physics: Conference Series. 2009;**149**:012109. DOI: 10.1088/1742-6596/149/1/012109

[9] Reindl M, Odenbach S. Physics of Fluids. 2011;**23**:093102. DOI: 10.1063/1.3633341

[10] Odenbach S, Müller HW. Journal of Magnetism and Magnetic Materials. 2005;**289**:242. DOI: 10.1016/j.jmmm.2004.11.069

[11] Mahle S, Ilg P, Liu M. Physical Review E. 2008;**77**:016305. DOI: 10.1103/PhysRevE.77.016305

[12] Müller HW, Liu M. Physical Review E. 2001;**64**:061405. DOI: 10.1103/PhysRevE.64.061405

[13] Debye PJW. Polar Molecules. New York: Dover; 1929 https://hdl.handle.net/2027/uc1.$b45989

[14] McTague JP. The Journal of Chemical Physics. 1969;**51**:133. DOI: 10.1063/1.1671697

[15] Martsenyuk MA, Raikher YL, Shliomis MI. Soviet Physics - JETP. 1974;**38**:413 http://www.jetp.ac.ru/cgi-bin/dn/e_038_02_0413.pdf

[16] Gilles de GP, Prost J, editors. The Physics of Liquid Crystals. Oxford: Clarendon; 1983. ISBN: 9780198517856

[17] Leschhorn A, Embs JP, Lücke M. Journal of Physics: Condensed Matter. 2006;**18**:S2633. DOI: 10.1088/0953-8984/18/38/S07

[18] Embs JP, May S, Wagner C, Kityk AV, Leschhorn A, Lücke M. Physical Review E. 2006;**73**:036302. DOI: 10.1103/PhysRevE.73.036302

[19] Taylor GI. Philosophical Transactions of the Royal Society of London A. 1923;**223**:289. DOI: 10.1098/rsta.1923.0008

[20] Chossat P, Iooss G. The Couette-Taylor Problem. Berlin: Springer; 1994. DOI: 10.1007/978-1-4612-4300-7

[21] Langevin P. Annales de Chimie Physique. 1905;**5**(70)

[22] Embs J, Müller HW, Wagner C, Knorr K, Lücke M. Physical Review E. 2000;**61**:R2196. DOI: 10.1103/PhysRevE.61.R2196

[23] Odenbach S, Gilly H. Journal of Magnetism and Magnetic Materials.

1995;**152**(123). DOI: 10.1016/0304-8853(95)00442-4

[24] Altmeyer S, Lopez JM, Do Y. Physical Review E. 2012;**85**:066314. DOI: 10.1103/PhysRevE.85.066314

[25] Zubarev AY, Skakova Y. Physical Review E. 2000;**61**:5415. DOI: 10.1103/PhysRevE.61.5415

[26] Golubitsky M, Stewart I, Schaeffer D. Singularities and Groups in Bifurcation Theory II. New York: Springer; 1988. ISBN: 978-1-4612-4574-2

[27] Golubitsky M, Stewart I. SIAM Journal on Mathematical Analysis. 1986;**17**:249. DOI: 10.1137/0517023

[28] Ch H, Altmeyer S, Pinter A, Lücke M. New Journal of Physics. 2009;**11**:053002. DOI: 10.1088/1367-2630/11/5/053002

[29] Davey A, Di Prima RC, Stuart JT. Journal of Fluid Mechanics. 1968;**31**:17. DOI: 10.1017/S0022112068000029

[30] Iooss G. Journal of Fluid Mechanics. 1986;**173**:273. DOI: 10.1017/S0022112086001179

[31] Altmeyer S, Do Y, Marques F, Lopez JM. Physical Review E. 2012;**86**:046316. DOI: 10.1103/PhysRevE.86.046316

[32] Coles D. Journal of Fluid Mechanics. 1965;**21**:385. DOI: 10.1017/S0022112065000241

[33] Wereley ST, Lueptow RM. Journal of Fluid Mechanics. 1998;**364**:59. DOI: 10.1017/S0022112098008969

[34] Jones CA. Journal of Fluid Mechanics. 1981;**102**:249. DOI: 10.1017/S0022112081002620

[35] Akonur A, Lueptow RM. Physics of Fluids. 2003;**15**:947. DOI: 10.1063/1.1556615

[36] King GP, Li Y, Lee W, Swinney HL, Marcus PS. Journal of Fluid Mechanics. 1984;**141**:365. DOI: 10.1017/S0022112084000896

[37] Altmeyer S, Lopez JM, Do Y. Physical Review E. 2013;**88**:013003. DOI: 10.1103/PhysRevE.88.013003

[38] Odenbach S, Müller HW. Physical Review Letters. 2002;**89**:037202. DOI: 10.1103/PhysRevLett.89.037202

[39] Müller HW, Hahn D, Liu M. Journal of Physics: Condensed Matter. 2006;**18**:S2623. DOI: 10.1088/0953-8984/18/38/S06

[40] Chang MH, Chen CK, Weng HC. International Journal of Engineering Science. 2003;**41**:103. DOI: 10.1016/S0020-7225(02)00183-0

[41] Singh J, Bajaj R. Journal of Magnetism and Magnetic Materials. 2005;**294**:53. DOI: 10.1016/j.jmmm.2004.10.123

[42] Singh J, Bajaj R. International Journal of Mathematics and Mathematical Sciences. 2005;**3727**. DOI: 10.1155/IJMMS.2005.3727

[43] Altmeyer S, Leschhorn A, Ch H, Lücke M. Physical Review E. 2013;**87**:053010. DOI: 10.1103/PhysRevE.87.053010

[44] Reindl M, Odenbach S. Experiments in Fluids. 2010;**50**:375. DOI: 10.1007/s00348-010-0940-y

[45] Hirt CW, Nichols BD, Romero NC. NASA STI/Recon Technical Report N; 1975. http://adsabs.harvard.edu/abs/1975STIN...7532418H

Chapter 3

Equilibrium Structures and Stationary Patterns on Magnetic Colloidal Fluids

Ricardo Peredo Ortíz, Martin Hernández Contreras and Raquel Hernández Gómez

Abstract

In this chapter, we review the experimental and theoretical modeling of structural and dynamical properties of colloidal magnetic fluids at equilibrium. Presently, several prototype experimental systems are very well characterized. We survey the different models, which help to reach a comprehensive knowledge of these complex magnetic fluids. One prime example is the ongoing investigation of the realistic interparticle potentials that drive the formation of the different phase states observed experimentally. Further, a stochastic equation approach for the description of tracer diffusion, viscoelasticity, and dielectric relaxation at equilibrium in colloidal ferrofluids is discussed.

Keywords: ferrofluids, viscoelasticity, colloidal magnetic fluids, magnetohydrodynamics, dielectric relaxation, diffusion stochastic dynamics

1. Introduction

Colloidal magnetic fluids are made up of nanometer and micron-size ferromagnetic particles dispersed either in electrolytes or in organic solvents [1]. Due to their easy manipulation with magnetic or electric fields, they are having a significant impact on diverse technological applications ranging from biomedicine [2] and photonic devices [3, 4] up to fundamental studies that motivate the development of modern theories of nonequilibrium condensed matter [5–10]. In this chapter, we review the studies of static structural and dynamical properties in colloidal magnetic fluids, which provide a comprehensive description of their bulk phase behavior at thermal equilibrium. Presently, model magnetic colloidal systems can be prepared with tunable interaction among particles of the hard sphere and long-range dipolar types [11–14], especially the colloidal system constituted of micron-size polystyrene spheres that are electrically charged and dispersed in organic solvents [11]. In such a system, particles acquire an induced electric dipole moment under the external static electric field. Confocal microscopy has allowed the determination of its three-dimensional bulk phase diagram as a function of applied electric field and by taking into account the presence of a 1:1 salt. The electric and magnetic dipole pair interaction potential is symmetric. Consequently, a corresponding change of physical units leads to an equivalent description of the magnetic fluid phase diagrams. Novel Monte Carlo simulation methods have

 IntechOpen

confirmed the different experimental phases [15, 16]. Integral equations and density functional theories have assisted in the understanding of the phases formed by molecular liquids [17], and they have been used to gain qualitative insight into the expected phases as a function of volume fraction and dipolar strength of colloidal magnetic fluids [18–23]. Likewise, the structure factor and diffusion coefficient of magnetic suspensions made of maghemite nanoparticles dispersed in water and equilibrium with electrolyte solutions have also been reported [24–31]. The measured structure factor provides the microscopic arrangements of particles accurately. Here, we provide a Langevin stochastic approach that allows the determination of the translational and rotational diffusion of the particles in ferrofluids [32, 33]. These dynamic properties, in turn, allow quantifying the viscoelastic and dielectric moduli of the structural evolution of the fluid toward their equilibrium states [34]. This approach may help to interpret recent experiments of passive microrheology [35, 36] and Altern Current spectroscopic techniques that measure the viscoelastic dynamics and dielectric relaxation of colloidal magnetic fluids. It is expected these theoretical methods be extended to help to understand the corresponding experimental observations of similar dynamics in two-dimensional paramagnetic colloids [37] and under external fields.

2. Experimental observations of structural properties of magnetic fluids

The importance of knowing the phase diagram of a ferrofluid resides on the information it provides about the undetermined underlying effective interaction of the ferromagnetic particles [13]. For maghemite nanoparticles in aqueous solutions, there have been attempts to determine such electrostatic potentials by using the measured parameters such as dipole strength and fluid density as inputs in Brownian dynamics simulations to reproduce the observed bulk structure factor [30, 31].

2.1 Importance of an experimental phase diagram of maghemite colloid

There is a set of experiments in a well-characterized system of electrically stabilized maghemite nanometer-size particles, which are dispersed in water [24–29]. Its stabilization is reached with the citrate electrolyte. By fine-tuning the salt concentration, the particle's interaction was shifted from repulsive, where they form a solid glass phase, into long-range attractive interactions that yield a fluid or gas state behavior. These experiments have prompted the determination of the corresponding pairwise interaction among particles [13]. The proposed interaction consists of a Yukawa repulsive part, a Van der Waals short-range attraction and includes an angular averaged attractive long-range pair dipole potential. There have not been attempts to predict the phase diagram with these potentials. However, the modeled interactions were used to determine with Brownian dynamics the observed structure factor [30, 31]. By using a generic model of pair dipole interactions together with an effective attraction associated with density gradients, Lacoste, et al. [38] considered a quasi-two-dimensional layer of a dilute ferrofluid subjected to a perpendicular magnetic field, and with the help of a mean field energy approach, they found modulated phases given by stripes and hexagonal formations. The Gibbs free energy they considered for inhomogeneous systems has the general shape $G = F_H - \mu\phi\left(\vec{r}\right)$, where F_H is the Helmholtz energy; the chemical potential μ constrains the average local composition $\phi\left(\vec{r}\right)$ or space-dependent volume fraction of particles at the position $\vec{r}$ to have a total fixed value $\int dr^2\phi\left(\vec{r}\right)$. Their model is a thin film of ferrofluid with a constant magnetic field acting perpendicular to the fluid layer. Since external fields produce an inhomogeneous distribution of particles, the volume fraction becomes a local function of position. F_H is constructed at mean field level to have the following terms: the internal energy of the pair dipole-dipole interactions $(1/2)\sum_{i,j(i\neq j)} V_{dd}\left(\vec{r}_{ij}\right)$ and the rotational free energy F_{rot} of an ideal gas of paramagnetic dipoles with polar angles as the only degrees of freedom under a constant magnetic field. F_{rot} is obtained from the partition function in the canonical ensemble calculated with the total energy of dipole $\hat{m}_i$ with field $\vec{H}$ interactions $-\sum_i \hat{m}_i\cdot\vec{H}$, the entropic energy contribution due to all possible spatial configurations of the finite size particles in the volume available $-k_BT[\phi\ln\phi+(1-\phi)\ln(1-\phi)]$ as given by the lattice gas. k_B is the Boltzmann constant and the room temperature T. The energy of particles that arise from gradients in their concentration and therefore setting their average available local volume is $(1/2)\int dr^2\left[\nabla\phi\left(\vec{r}\right)\right]^2$, which is provided by a continuous version of the lattice gas model. The ferroparticles are all equal, and their energy of pairwise interaction for two dipoles separated with the center-to-center vector distance $\vec{r}$ is given by $V_{dd}=-\left(m^2/r_{ij}{}^4\right)\left[3\vec{r_{ij}}\cdot\hat{u}_i\vec{r_{ij}}\cdot\hat{u}_j-\hat{u}_i\cdot\hat{u}_i\right]$, where m is the particle's constant dipole moment, and $\hat{u}_i$ is the unitary vector orientation of particle i. Thus, the resulting mean Gibbs energy G is mapped from the lattice gas model into a continuous position dependence representation. The free energy is a function of the local dipole moment and concentration $m(r)$, $\phi(r)$. A ferrofluid in a continuous single phase such as a dispersed homogenous ferrofluid is an equilibrium state with an average concentration $\overline{\phi}$ and bulk dipole moment $\overline{m}$. This state is characterized by the minimum of the free energy, that is, $\partial G/\partial m=0$, $\partial G/\partial\phi=0$, which provides the average values of bulk dipole moment and concentration above. However, it is observed that for specific magnitudes of the external field and ferrofluid concentration, there appears chaining of particles that coalescence into sheets, which at higher field strength evolve to labyrinthine structures. The nucleation of these structures starts with the formation of magnetization domains associated with small clustering of particles with sizes on the order of a few microns. Short-wavelength thermal fluctuations of this size are capable of describing the formation of nonuniform phases. One way to analyze the formation of these phases uses the expansion of the free energy difference [38] $\Delta G(\varphi,m)=G(\varphi,m)-G(\overline{\varphi},\overline{m})$ up to second order in the fluctuations of dipole magnetization $\delta m=m(r)-\overline{m}$, and concentration $\delta\phi=\phi(r)-\overline{\phi}$ about their mean values. In the two-dimensional Fourier space representation, the free energy difference is a quadratic form $\Delta G=(1/2\pi^2)$ $\int d^2\vec{k}\left[a\left|\delta m\left(\vec{k}\right)\right|^2+b\delta m\left(\vec{k}\right)\delta\phi\left(-\vec{k}\right)+c\left|\delta\phi\left(\vec{k}\right)\right|^2\right]$ with a, b, c related to the material parameters of ferrofluid external field $\overline{m}$ and $\overline{\phi}$. The uniform phase becomes unstable to the formation of modulated nonuniform phases if the determinant of the integrand gets negative at the minimum of wave number [38].

It has been recognized that ferromagnetic particles of 10 -nm size remain dispersed [39]; however, when particles have larger sizes up to the micron scale, experiments find they do assemble into chains, rings, and nets were several chains bound together into an amorphous structure [40]. A mean field theory that takes into account such topological structures was developed to explain the liquid-gas phase transition by the formation of Y-like bounds of chains in the network [41]. There remains a quantitative verification of this theory with its experimental counterpart and with computer simulations. Another method to validate the proposed model of particle's interaction is based on the parameters that appear in the Lennard-Jones potential that represents the Van der Waals attractions together with the pair dipole-dipole potential, and by fitting the predicted magnetization curve as a function of the applied magnetic field to the experimental one [42]. Even though there is not yet a realistic model interaction potential that represents the interaction of ferromagnetic particles, the use of fitting parameters as mentioned above has led to the prediction of the structure factor and the birefringence as a function of the applied field, which shows qualitative agreement with the experimental values [22, 29–31].

2.2 Experimental phase diagram of a charged and sterically stabilized electrorheological colloidal suspension under an electric field

Yethiraj et al. [11] made a colloidal system of charged and sterically stabilized polymethyl methacrylate spheres of micron size in an organic solvent to have dipole moments on particles induced by an electric field. This monodisperse suspension shows both long-range repulsion, and attractive anisotropic interaction potential that is fixed with the addition of salt, whereas the dipolar interaction is controlled by the external electric field. For low Reynolds numbers and in an infinite fluid, a solid spherical particle that sinks in the fluid reaches a terminal velocity due to a balance of the friction force exerted by the fluid that opposes the gravitational force on the particle. If the fluid has a dynamic viscosity η and mass density ρ_f, and the spherical particle possesses a mass density ρ_m, then the steady velocity is $V = \left(\rho_f - \rho_m\right)gd^2/18\eta$, where g is the gravitational acceleration and d the particle diameter. Thus, the effect of gravity is diminished by matching the density of the particles to that of the solvent. The Van der Waals attraction that otherwise would produce aggregation was reduced by matching the index of refraction to the visible. The Van der Waals attractive energy of two spherical particles is $V_{vdw}(r) = -(A/12)\left[d^2/\left(r^2 - d^2\right) + d^2/r^2 + 2\ln\left(1 - d^2/r^2\right)\right]$, where r is the separation distance of the spheres' centers [43]. Therefore, the Hamaker constant [44] $A = 3h\nu\left(n_p - n_s\right)^2 \left(n_p + n_s\right)^2/16\sqrt{2}\left({n_p}^2 + {n_s}^2\right)^{3/2}$ can be made zero if the optical refractive index of the colloidal particles n_p is equal to that of the solvent n_s for an appropriated selection of the frequency ν of the visible light. Here, h is the Planck constant. With confocal microscopy studies, they observed in real-space representation crystalline structures controlled by the repulsive part of the potential, which induces a structure of random hexagonal close packing. For longer-range repulsion, it becomes face-centered cubic (fcc), and one can also find a body-centered cubic phase (bcc). As a function of the electric field, a dipolar phase diagram was established. For low field, there is a sequence of fluid-bcc-fcc (phase-centered cubic) phases. For larger values of the field, there is string formation that coalesces into sheets that in turn coarsen to form body-centered tetragonal crystallites. Moreover, for increasing volume fractions, the string fluid crystallizes to a space-filling tetragonal phase. Such a rich variety of phases can be generated controllably by the concentration of the electrolyte. High salt concentration screens the electrostatic repulsion becoming short-distance hard sphere interaction whereas the small content of salt increases the spatial range of the potential. The dipolar interaction is controlled by the external electric field. Consequently, due to the symmetric form of the electric and magnetic dipolar potential, a simple change of physical units leads to the similar conclusions to be expected for a colloidal magnetic ferrofluid. For oily ferrofluids, the short-range Van der Waals attraction and hardcore repulsion are modeled by the

Lennard-Jones interaction energy $V_{LJ}(r) = 4\varepsilon_0\left[(d/r)^{12} - (d/r)^6\right]$ with ε_0 the potential depth that can be estimated experimentally. For ferrofluid in an aqueous solvent of dielectric constant ε and monodisperse spherical particles, due to the presence of salt at density ρ_i, the Yukawa potential $V_{yuk}(r) = Q^* \exp\left[-\kappa(r-d)\right]/r$, $\kappa^2 := 4\pi e^2\rho_i/\varepsilon k_B T$ is used instead, where e is the electron charge. Q^* is the effective electric charge of a particle that can be measured with electrophoresis experiments.

2.3 Theoretical models of phase diagrams of dipolar colloidal fluids: computer simulation and optimization techniques

Finite-temperature calculations of ground state free energies with Yukawa repulsion and dipolar-dipolar interactions of varying strength were performed by Hynninen and Dijkstra [15]. In their calculations, they used canonical Monte Carlos simulations to obtain the free energy minimum as a function of volume fraction and dipolar strength in three-dimensional model systems. They found a phase diagram that contains the same phases as was observed by Yethiraj et al. [11]. However, additionally, they predicted a new hexagonal close-packed phase at the hard sphere repulsion limit and body-centered orthorhombic when the repulsion becomes long range by lowering the content of electrolyte in solution. Their method allows quicker and reliable determinations of the three-dimensional phase diagram than the evolutionary algorithm method that searches for the ground state at zero temperature for two-dimensional systems [45]. The bulk phase diagram has not been searched yet with genetic algorithm (GA) at finite temperature. A particular detail of GA is that to look for the minimum of the free energy, the derivative of the energy function at each evolution step is required for finding the best-adapted crystal structures (individuals) in the population, which is made up of several crystal structures produced by the algorithm. The search of the best-adapted individuals (crystal phase) is done in the energy landscape. For fixed pressure and finite temperature, the Gibbs free energy as a function of pairwise interaction strength and particles concentration in the fluid yields a three-dimensional plot of the free energy known as energy landscape, which has local minima and maxima. The global minimum of an equilibrium crystal structure is reached with the assistance of a local gradient algorithm that requires the derivative of the energy. Such a procedure makes an indirect search of the crystal structure by sampling the energy landscape. Thus, it becomes technically more involved to implement than the Hynninen et al. method of MC ground state energy searches. In **Figure 1**, we provide two nonequilibrium states of local minima that result from the application of a genetic algorithm approach at zero temperature and fixed pressure.

Recently, Spiteri and Messina [16] have proposed an efficient nonlinear optimization technique that allows predicting the crystal phases of dipolar colloids at zero temperature without the need to use derivatives of the free energies. It was found that for a monodisperse repulsive hard sphere plus dipole-dipole interaction, the predicted phase diagram has the same phases as was previously found by Hynninen et al. [15] and that a knew so-called clinohexagonal prism span all known ground state structures at any density. There remains to be verified if this prediction fulfills at a finite temperature also.

2.4 Experimental structure factors and diffusion coefficients

The microstructural order inside of the ferrofluid is determined by the potential interaction among particles. The measured structure factor yields the details of the particles' spatial geometric arrangement. The water-based maghemite ferrofluid is

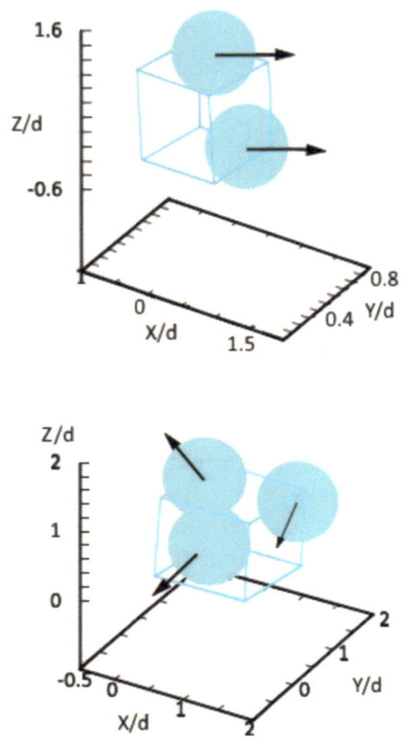

Figure 1.
Genetic algorithm prediction of minimum local structures at two densities of equally sized ferroparticles in the aqueous colloidal magnetic fluid at $T = 0K$ and constant pressure. X, Y, and Z are Cartesian coordinates. Both structures are simple cubic cells. The bottom structure has the highest density and the lowest enthalpy energy. Black arrows depict the direction of the particle's dipole moment.

electrostatically stabilized. The interparticle interaction was modeled in Ref. [13] with a Yukawa repulsion part together with the anisotropic dipolar potential. The highest peak of the structure factor regarding wave number k, as a function of applied field, shows the same qualitative tendency as the experimental peak. For the same type of pair potential, a calculation of the structure factor using the rescaled mean spherical approximation of liquid theory was developed by Wagner et al. [46]. Their calculated structure factor compares well with the measured one in a cobalt-based aqueous colloid. Their fabricated ferrofluid has a polydispersity in particle sizes, which were successfully taken into account via a Shultz distribution. The microstructure in the suspension has also been modeled by Pyanzina et al. [47] with an analytical pair correlation function obtained through a thermodynamic expansion of the density and dipole strength. They find excellent agreement between the theoretical calculation and molecular dynamics data for the scattering vector dependence of the pair correlation at the dilute limit. The agreement improves at low volume fraction whereas at higher concentrations, the coincidence occurs only about the main first peak. Recently, we determined with Langevin dynamics simulations the structure factor in model monodisperse ferrofluids of spherical particles that interact through Lennard-Jones plus dipolar pair potentials. As previously found [17, 42], we also find that at the low colloid density and dipole strength, the particles are well dispersed; however, for higher dipole moment, there is chaining of particles. If the concentration rises, then the system shows a small cluster with few particles as can be seen in **Figure 2** of the structure factor $S(k)$.

The contact values at k = 0 yield the compressibility modulus. The material parameters of maghemite colloid were used [40]. At low densities and magnetic moment, there is liquid order. For higher dipole moments, formation of chain appears.

Figure 3 is a plot of the collective diffusion coefficient D_C normalized to the free diffusion of a single particle D^0 in the hydrodynamic limit of zero wave number. It is given as $D_c = D^0/S(k=0)$. In the left-hand side plot of **Figure 3(a)** with symbol ∘ the collective diffusion coefficient increases as the magnitude of dipole moment increases at high-density limit of $\rho^* = 0.9$. However, a reverse behavior is shown by this collective diffusion property when the density is very low $\rho^* = \rho d^3 = 0.1$ (plot with a symbol •, $\rho = N/V$, N the number of particles, and V the volume of the fluid). We should note that in the former case, there occurs the formation of chains, and at the highest concentration, the particles are dispersed.

The researchers in Ref. [46] also measured the wave vector-dependent collective translational diffusion coefficient in the absence of an external field using X-ray

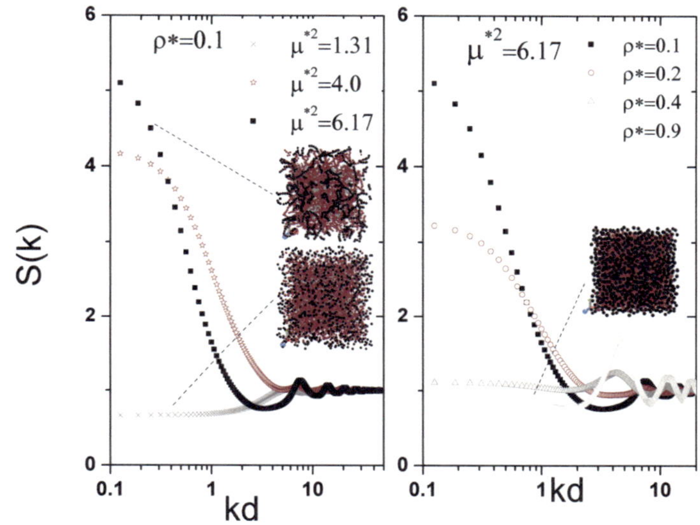

Figure 2.
Simulation results of structure factor versus wave number. The contact values at k=0 yield the compressibility modulus. The material parameters of maghemite colloid were used [40]. At low densities and magnetic moment, there is liquid order. For higher dipole moments formation of chain appears.

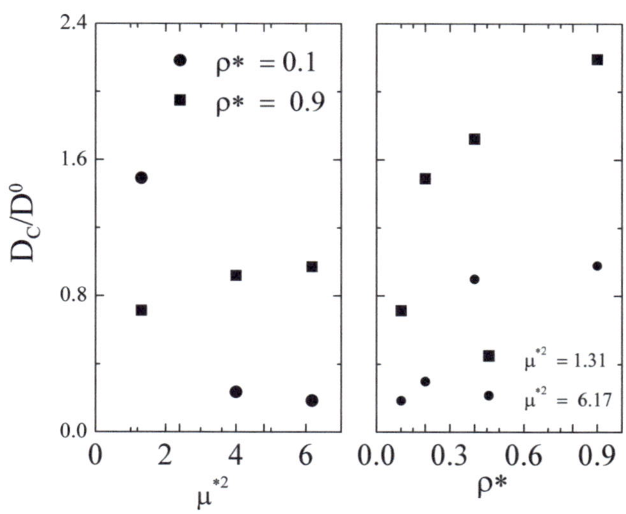

Figure 3.
Calculated collective diffusion coefficient D_c normalized to the free particle diffusion constant at the hydrodynamic limit of low wavenumber k = 0. Notice that at high density, $\rho^* = 0.9$, this property increases (left, black square symbol).

correlation spectroscopy. Such a dynamical property has a known linear dependence on the particles' hydrodynamic interactions (HIs). The proposed theoretical hydrodynamic function that captures the HI among particles overestimates the experimental data. On the other hand, Meriguet et al. [29] used small-angle neutron scattering to measure the incoherent scattering function at the fixed magnetic field. The incoherent part gives in the long-time overdamped regime, the single translational diffusion coefficient for long wave vectors k, and the coherent part provides the collective diffusion as a function of k. Using liquid theory definitions of these two dynamical functions, their comparison with the experimental values yields good agreement at all wave scattering vectors k. The same collective diffusion coefficient can be measured by forced Rayleigh scattering also [26]. In this technique, the fluctuations in concentrations of the particles are considered a dynamic variable. We note that similar studies have been undertaken in the case of rodlike macromolecules in suspension such as Tobacco mosaic virus [48]. Explicit statistical mechanics derivations for the dynamical collective correlation functions (intermediate scattering function) have been given that fit well Brownian dynamics simulations just at long wave numbers but fail at intermediate and short k values. In Refs. [32–34], we proposed a Langevin equation theory for a tracer ferroparticle whose motion is coupled to the cloud of the other colloidal magnetic particles with which it interacts.

$$\begin{aligned} M\frac{dV(t)}{dt} &= -\zeta^0 \cdot V(t) - \zeta^0_{TR} \cdot W(t) + f^0(t) + F(t), \\ I \cdot \frac{dW(t)}{dt} &= -\zeta^0_R \cdot W(t) - \zeta^0_{RT} \cdot V(t) + t^0(t) + T(t). \end{aligned} \tag{1}$$

V, W are the translational and angular velocities of the tracer particle and referred to a space-fixed frame at the center of mass of the tracer and following the orientation of the main tracer axis of symmetry. M and I are the mass and particle's matrix of the moment of inertia, respectively. There are neither hydrodynamic interactions among particles nor external magnetic fields. The first two terms in Eq. (1) are the solvent friction force and torque. The short-time free particle diagonal friction tensors ζ^0, ζ^0_{TR}, ζ^0_{RT}, ζ^0_R represent hydrodynamic drag forces and torques. These friction and random forces are the only quantities that convey information on the nature of the solvent. They need to be measured experimentally or provided by a free parameter. Moreover, they ignore the molecular degrees of freedom of position and orientation coordinates and momenta. They are coupled to the thermally driven solvent random forces f^0 and torques t^0 by fluctuation-dissipation theorems [32–34]. The total force and torque F, T on the tracer by the rest of the colloidal particles, which are at the concentration $n\left(\vec{r},\Omega,t\right) = \sum_{i=1}^{N}\delta\left(\vec{r}-\vec{r}_i(t)\right)\delta(\Omega-\Omega_i(t))$, are given by

$$\begin{aligned} F(t) &= \int d^3\vec{r}\, d\Omega \nabla\psi\left(\vec{r},\Omega\right)n\left(\vec{r},\Omega,t\right), \\ T(t) &= \int d^3\vec{r}\, d\Omega\left[\vec{r}\times\nabla+\nabla_\Omega\right]\psi\left(\vec{r},\Omega\right)n\left(\vec{r},\Omega,t\right) \end{aligned} \tag{2}$$

with $\Omega = (\theta,\varphi)$ being the polar angles, $\psi\left(\vec{r},\Omega\right) = V_{LJ} + V_{dd}(\mu_0/4\pi)$ is the pair potential with μ_0 the magnetic permeability of vacuum, and $\nabla_\Omega = \hat{u}\times d/d\hat{u}$ the angular gradient operator. The total force and torque are calculated using the probe's dipole located along the Z axis direction of its local frame, that is $\Omega = (0,0)$. The unitary Cartesian vector $\hat{u}$ of the orientation of any other particle is in the direction of that particle's axis of symmetry. The Langevin equation is coupled to the diffusion equation of the concentration of particles around the tracer through the total force on the tracer, which is linear in the instantaneous concentration of particles.

$$\overleftrightarrow{M}\cdot\frac{d\overleftrightarrow{V}(t)}{dt}=-\overleftrightarrow{\zeta}^{0}\cdot\overleftrightarrow{V}(t)+\overleftrightarrow{f}(t)+\int d^{3}\vec{r}d\Omega\overleftrightarrow{\nabla}\left[\psi\left(\vec{r},\Omega\right)\right]n\left(\vec{r},\Omega,t\right)\tag{3}$$

Here, $\overleftrightarrow{V}=(V,W)$, $\overleftrightarrow{\nabla}=\left(\nabla,\vec{r}\times\nabla+\nabla_{\Omega}\right)$. and the diffusion equation of the concentration $\partial n\left(\vec{r},\Omega,\mathrm{t}\right)/\partial t=-\nabla\cdot\left[\overleftrightarrow{V}n(t)\right]$ can be linearized altogether with Eq. (3) in $\delta\mathrm{n(t)}=\mathrm{n(t)}\text{-}\mathrm{n}^{eq}\left(\vec{\mathrm{r}},\Omega\right)$ around the local equilibrium concentration of particles $\mathrm{n}^{eq}\left(\vec{\mathrm{r}},\Omega\right):=\left\langle n\left(\vec{r},\Omega,t\right)\right\rangle$. Thus, in the center of mass of the tracer, and using general results of linear irreversible thermodynamics, the diffusion equation reads

$$\begin{aligned}\frac{\partial\delta n\left(\vec{r},\Omega,t\right)}{\partial t}=&\left[\overleftrightarrow{\nabla}n^{eq}\left(\vec{r},\Omega\right)\right]\cdot\overleftrightarrow{V}-\int_{0}^{t}dt'\int d^{3}\vec{r}'d\Omega'\int dr''d\Omega''L\left(\vec{r},\vec{r}',\Omega,\Omega';t-t'\right)\\&\times\sigma^{-1}\left(\vec{r}',\vec{r}'',\Omega',\Omega''\right)\delta n\left(\vec{r}'',\Omega'',t'\right)+\overleftrightarrow{\nabla}\cdot j\left(\vec{r},\Omega,t\right).\end{aligned}\tag{4}$$

where the Onsager coefficient $L=\left\langle\overleftrightarrow{\nabla}\cdot j\left(\vec{r},\Omega,t\right)\cdot\overleftrightarrow{\nabla}\cdot j^{\dagger}\left(\vec{r},\Omega,t\right)\right\rangle$ is related to the spatial variation of the stochastic current $j\left(\vec{r},\Omega,t\right)$. † means transpose and complex conjugate (adjoint operator). The inverse of the static correlation function σ fulfills a relationship with the potential and concentration function [32–34]. Using methods of Green functions Eq. (4), one solves for $\delta n\left(\vec{r},\Omega,t\right)$, and the result is further substituted into the linearized version in the concentration of Eq. (3) yielding the tracer Langevin equation

$$\overleftrightarrow{M}\cdot\frac{d\overleftrightarrow{V}(t)}{dt}=-\overleftrightarrow{\zeta}^{0}\cdot\overleftrightarrow{V}(t)+\overleftrightarrow{f}(t)-\int_{0}^{t}dt'\Delta\overleftrightarrow{\zeta}(t-t')\cdot\overleftrightarrow{V}(t)+\overleftrightarrow{F}(t)\tag{5}$$

$\overleftrightarrow{F}$ are fluctuating forces arising from the spontaneous departure from zero of the net force and torque on the tracer. It satisfies the fluctuation-dissipation relationship with the tracer-effective friction coefficient $\left\langle\overleftrightarrow{F}(t)\overleftrightarrow{F}^{\dagger}(0)\right\rangle=k_{B}T\Delta\overleftrightarrow{\zeta}(t)$. $T=300K$ is the room temperature. As a result, explicit expressions for the tracer translational and rotational friction coefficients are given regarding the equilibrium concentration $\mathrm{n}^{eq}\left(\vec{\mathrm{r}},\Omega\right)$ that embodies the microstructure of the particles inside of the liquid, and the direct pair potential ψ,

$$\begin{aligned}\Delta\overleftrightarrow{\zeta}(t)=&\frac{k_{B}T}{(4\pi)^{2}}\int d^{3}\vec{r}d^{3}\vec{r}'d^{3}\vec{r}''d\Omega d\Omega'd\Omega''\left[\overleftrightarrow{\nabla}n^{eq}\left(\vec{r},\Omega\right)\right]\\&\times\sigma^{-1}\left(\vec{r},\vec{r}',\Omega,\Omega'\right)\chi\left(\vec{r}',\vec{r}'',\Omega',\Omega'',t\right)\left[\overleftrightarrow{\nabla}''n^{eq}\left(\vec{r}'',\Omega''\right)\right]^{\dagger}.\\\Delta\overleftrightarrow{\zeta}(t)=&\frac{1}{k_{B}T(4\pi)^{2}}\int d^{3}\vec{r}d^{3}\vec{r}'d^{3}\vec{r}''d\Omega d\Omega'd\Omega''\left[\overleftrightarrow{\nabla}\psi\left(\vec{r},\Omega\right)\right]\\&\times\chi\left(\vec{r},\vec{r}',\Omega,\Omega',t\right)\sigma\left(\vec{r}',\vec{r}'',\Omega',\Omega''\right)\left[\overleftrightarrow{\nabla}''n^{eq}\left(\vec{r}'',\Omega''\right)\right]^{\dagger}\end{aligned}\tag{6}$$

The propagator χ is obtained from the diffusion Eq. (4), and it is related to the van Hove function $C\left(\vec{r},\Omega,\vec{r}',\Omega',t\right) := \left\langle \delta n\left(\vec{r},\Omega,t\right)\delta n\left(\overleftarrow{r}'\Omega' 0\right)\right\rangle = \int d^3\vec{r}''d\Omega''$ $\chi\left(\vec{r},\vec{r}'',\Omega,\Omega'',t\right)\sigma\left(\vec{r}'',\vec{r}',\Omega'',\Omega'\right)$ whose dynamical evolution is attained from Eq. (4) after multiplying both sides by $\delta n\left(\vec{r},\Omega,t=0\right)$ and taking further an ensemble average. The resulting dynamical equation of the van Hove function has at t = 0 the initial condition $\int d^3\vec{r}''d\Omega''\sigma\left(\vec{r},\vec{r}'',\Omega,\Omega''\right)\chi\left(\vec{r}'',\vec{r}',\Omega'',\Omega',t=0\right) =$ $\sigma\left(\vec{r},\vec{r}',\Omega,\Omega'\right)$. The properties in Eq. (6) depend on the knowledge of the structure factor of the ferrofluid, which takes into account the micro-structural order of particles inside the ferrofluid as dictated by the underlying interaction potential. It is valid at high concentrations but ignores hydrodynamic interactions among particles. For monodisperse colloids, the tracer is equal to the other particles and the fluid is homogeneous. Thus, the distribution of particles is $n^{eq}(r,\Omega) =$ $\rho g\left(r=\left|\vec{r}\right|,\Omega,\Omega'(0,0)\right)$. Notice that the pair correlation function is defined between the tracer orientation $\Omega'(0,0)$, and another particle with $\Omega(\theta,\varphi)$, separated by the mean distance $r = \left|\vec{r}-\vec{r}'\right|$. Due to the spatial symmetry of the potential ψ, the pair correlation function has the rotationally invariant expansion $g(r,\Omega,\Omega') =$ $g(r) + h_\Delta(r)\Delta + h_D(r)D_1$, with $\Delta := \hat{u}_1\cdot\hat{u}_2$, $D_1 := \left[3\hat{\vec{r}}_{12}\cdot\hat{u}_1\hat{\vec{r}}_{12}\cdot\hat{u}_2 - \hat{u}_2\cdot\hat{u}_1\right]$, where $\hat{\vec{r}} = \vec{r}/r$, $\hat{u}$ is the unitary dipole vector with orientation $\hat{u} = \hat{u}(\theta,\varphi)$. Using the Fick approximation for the Onsager coefficient, $L(r,r',\Omega,\Omega') = \rho\left[D^0\nabla^2 + D_R^0\nabla_\Omega\right]$ $\delta(r-r')\delta(\Omega-\Omega')$, the unknown propagator in Eq. (4) $\chi = \exp\left(-tL\circ\sigma^{-1}\right)$, $\circ :=$ $\int d^3\,\vec{r}\,d\Omega$ can be fully determined. For a spherical particle, the free translational and rotational diffusion coefficients are $D^0 = k_BT/\zeta^0$, $D_R^0 = k_BT/\zeta_R^0$, with $\zeta^0 = 3\pi\eta_{sol}d$, $\zeta_R^0 = \pi\eta_{sol}d^3$, and η_{sol} the solvent viscosity. Taking the Laplace transform $\Delta\zeta(\omega) =$ $\int_0^\infty dte^{i\omega t}\Delta\zeta(t)$ of the friction tensor in Eq. (6), it yields the real part of the translational and rotational diffusion coefficients of the tracer at the longtime overdamped regime $D = k_BT/\left[\zeta^0 + \mathrm{Re}\Delta\zeta(\omega=0)\right]$, $D_R = k_BT/\left[\zeta_R^0 + \mathrm{Re}\Delta\zeta_R(\omega=0)\right]$ [34], where

$$\begin{aligned}\mathrm{Re}\Delta\zeta &= \frac{\zeta^0}{6\pi^2\rho^*}\int_0^\infty dxx^4\frac{\left(S_{,0}^{00}(x)-1\right)^2\tau_{s,\,iso}^*}{S_{,0}^{00}(x)}\\&\quad+\frac{128\pi\zeta^0\rho^*\mu^{*4}}{3}\int_0^\infty dxj_2(x)^2\left(6+x^2\right)\left[2\tau_{s,1}^{*2}+3\tau_{s,0}^{*2}\right],\\\mathrm{Re}\Delta\zeta_R &= \frac{128\pi\rho^*\mu^{*4}\zeta_R^0}{105}\int_0^\infty dxj_1(x)^2\left(6+x^2\right)\left[67\tau_{s,1}^{*2}+38\tau_{s,0}^{*2}\right],\end{aligned} \tag{7}$$

j_l is the spherical Bessel function of order l, $\tau_{s,\,iso}^* = 1/\left[x^2\left(1+1/S_{,0}^{00}(x)\right)\right]$, $\tau_{s,\,\alpha}^* = S_{,\alpha}^{11}(x)/[8\pi(x^2+6)]$, $x := kd$. The structure factor components of the dipolar liquid are

$$\begin{aligned}S_{,0}^{00}(x) &= 1+\frac{\rho^*}{d^3}[g(x)-1],\ S_{,0}^{11}(x) = 1+\frac{\rho^*}{3d^3}[h_\Delta(x)+2h_D(x)],\\S_{,1}^{11}(x) &= 1+\frac{\rho^*}{3d^3}[h_\Delta(x)-h_D(x)].\end{aligned} \tag{8}$$

These correlation function components are obtained from the Fourier transform of their definitions

$$g(r) = \frac{\left\langle \sum_{i,\, j(i \neq j)} \delta\left(r - \left|\vec{r}_{ij}\right|\right)\right\rangle}{N 4\pi\rho^* r^2}, h_D(r) = \frac{3}{2}\frac{\left\langle \sum_{i,\, j(i \neq j)} \delta\left(r - \left|\vec{r}_{ij}\right|\right)\left[3\hat{\vec{r}}_{ij}\cdot\hat{u}_i\hat{\vec{r}}_{ij}\cdot\hat{u}_j - \hat{u}_i\cdot\hat{u}_j\right]\right\rangle}{N 4\pi\rho^* r^2},$$

$$h_\Delta(r) = 3\frac{\left\langle \sum_{i,\, j(i \neq j)} \delta\left(r - \left|\vec{r}_{ij}\right|\right)\hat{u}_i\cdot\hat{u}_j\right\rangle}{N 4\pi\rho^* r^2} \tag{9}$$

In **Figure 4,** Ref. [34] is provided a plot of the translational and rotational self-diffusion coefficients of such a theory. In this picture, we compare the diffusion coefficients with Langevin dynamics simulation results for the same diffusion coefficients by using typical parameters of $\mu^{*2} = \mu_0 m^2/\left(4\pi d^3 k_B T\right)$ for the real ferrofluid made of Fe_2O_3. μ_0 is the magnetic permeability of vacuum. The material data are: [40] $\varepsilon_0 = 5.45311 \times 10^{-23} J, d = 10^{-8} m$, particle mass and the viscosity are $m_0 = 2.70710 Kg, \eta_{sol} = 0.852 \times 10^{-3} Kg/ms$, respectively.

2.5 Diffusion microrheology of colloidal magnetic fluids under no external electric or magnetic fields

Ferrofluids are complex fluids whose viscoelastic response to weekly applied strain rates $\dot{\gamma}(t)$ is represented by a constitutive equation for the shear stress $\sigma(t) = \int_0^t dt' G(t-t')\dot{\gamma}(t)$ where the dot means time derivative and the shear modulus is $G(t)$. In the representation of frequency space ω, the shear is $\sigma(\omega) = G(\omega) i\omega\gamma(\omega)$. Mason and Weitz [49] assumed that the macroscopic temporal relaxation of $G(t)$ is the same time scale of the viscosity that affects a tracer particle that diffuses in the viscoelastic fluid. Because the shear stress divided by the strain rate has dimensions of viscosity, Mason et al. [49] defined the viscosity as $\eta(t) = \sigma(t)/\dot{\gamma}(t)$. Thus, they proposed that for spherical particles undergoing translational diffusion and in the frequency space, it is $G(\omega) = i\omega\eta(\omega) = i\omega\zeta(\omega)/3\pi d$ for a particle experiencing

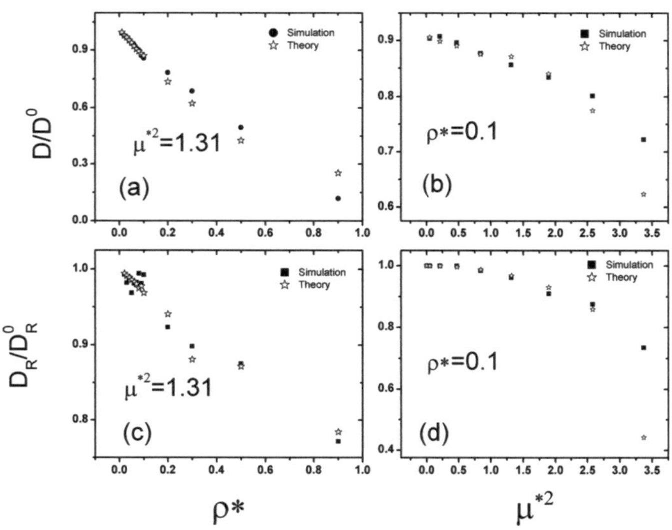

Figure 4.
Translational (rotational) self-diffusion coefficients D/D^0 $\left(D_R/D_R^0\right)$ versus density (a) ((c)), and dipole strength (b) ((d)). Simulation calculations are depicted with a black symbol and theory of Eq. (6) with ∘.

effective friction $\zeta(\omega)$. Here, we propose under the same conditions that the complex shear modulus of a ferrofluid is given by $G(\omega) = G' + iG'' = i\omega(\zeta^0 + \Delta\zeta(\omega))/3\pi d = i\omega\eta(\omega)$. Since the complex viscosity of a viscoelastic media is defined by $\eta = \eta' - i\eta'$, we find for the elastic and dissipative moduli, respectively [34]

$$G'(w) = \omega\eta''(\omega) = \int_0^\infty dx \left[\frac{P_0^{00}(x)\left(\omega^*\tau^*_{s,\,iso}\right)^2}{1+\left(\omega^*\tau^*_{s,\,iso}\right)^2} + \frac{P_1^{11}(x)\left(\omega^*\tau^*_{s,\,1}\right)^2}{1+\left(\omega^*\tau^*_{s,\,1}\right)^2} + \frac{P_0^{11}(x)\left(\omega^*\tau^*_{s,\,0}\right)^2}{1+\left(\omega^*\tau^*_{s,\,0}\right)^2} \right],$$

$$G''(w) = \omega\eta'(\omega) = \frac{k_B T\omega^*}{3\pi d^3} + \int_0^\infty dx \left[\frac{P_0^{00}(x)\omega^*\tau^*_{s,\,iso}}{1+\left(\omega^*\tau^*_{s,\,iso}\right)^2} + \frac{P_1^{11}(x)\omega^*\tau^*_{s,\,1}}{1+\left(\omega^*\tau^*_{s,\,1}\right)^2} + \frac{P_0^{11}(x)\omega^*\tau^*_{s,\,0}}{1+\left(\omega^*\tau^*_{s,\,0}\right)^2} \right],$$

$$P_0^{00}(x) = \frac{k_B T x^4\left(S^{00}_{,0}(x)-1\right)^2}{18\pi^3 d^3\rho^* S^{00}_{,0}(x)},\ P_1^{11}(x) = -\frac{32k_B T\rho^*\mu^{*4}j_2(x)^2 S^{11}_{,1}(x)}{9\pi d^3},$$

$$P_0^{11}(x) = \frac{48k_B T\rho^*\mu^{*4}j_2(x)^2 S^{11}_{,0}(x)}{9\pi d^3}, \tag{10}$$

where $\omega^* = \omega t_B$, $t_B = d^2/D^0$. **Figure 5** provides a comparison of the viscoelastic moduli of Eq. (10) with Langevin dynamics simulations for the same parameter as those of **Figure 4**.

We notice that the effective viscosity at the overdamped regime $\omega = 0$, which results from Eq. (10), is $\eta_0 = t_B G_o$, where the storage modulus of the ferrofluid is $G_o = \left(k_B T/3\pi d^3\right)D^0/D$. At low volume fraction $\phi = \pi\rho^*/6$ of colloidal particles in the magnetic fluid, we found the following expression of the viscosity $\eta_0 \approx \eta_{sol}(1 + 5\phi/2)$, which is the known exact Einstein result for low volume fractions if we assume that the free parameter ζ^0 can be replaced by the approximated form proposed by Mazur and Geigenmüller $\zeta^0(1 + 1.5\phi)/(1 - \phi)$ [50], which describes very well the free particle diffusion coefficient for all volume fractions.

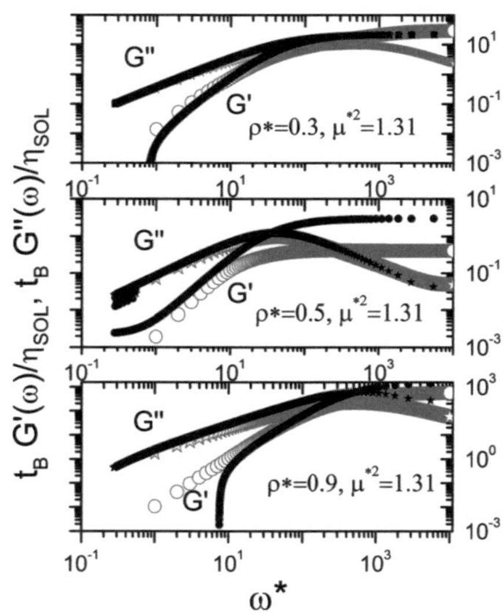

Figure 5.
Logarithmic plot of elastic G' and loss G'' moduli as a function of dimensionless frequency ω^* at three different concentrations and fixed dipole moment. Theory for $G'(G'')$ is denoted by $\circ(*)$ and that with simulation by $\bullet$, $*$.

2.6 Dynamic susceptibility response of a ferrofluid

The time-dependent magnetic susceptibility $\chi(t)$ of a ferrofluid can be measured either under a time-oscillating external magnetic field or in its absence with Altern Current dielectric spectroscopic techniques [51]. Our approach to calculating this property is based on the diffusion equation Eq. (4) that leads us to an expression for the collective dynamical property $\chi(t)$ regarding the friction Eq. (6). For this purpose, we multiplied Eq. (4) by $\delta n(t=0)\sum_{i,j}\hat{u}_i(t)\hat{u}_j(t)$, then we made a statistical average and Fourier transformed to write the dynamical equation of the collective magnetic tensor function

$$\langle M_k(t)M_{-k}(t=0)\rangle = -\left(m^2\mu_0 N/12\pi\right)C^{11}_{,1}(k,t), M_k(t) := \left(\mu\sqrt{\mu_0}/(4\pi)^{3/2}\right)$$
$$\times \sum_{i=1}^{N}\int d^3\vec{r} e^{ik\cdot\vec{r}_i}(t)d\Omega\hat{u}_x(t)\delta n(r,\Omega,t), \tag{11}$$

where $C^{11}_{,1}$ is the transversal component of the van Hove function. We extended the hydrodynamic approach of Hess and Klein [52] to include rotational Brownian motion of the particles and derived the dynamical equation of the Van Hove function

$$\frac{\partial C(k,\Omega,t)}{\partial t} = \int_0^t dt' d\Omega' M(k,\Omega,\Omega',t-t')C(k,\Omega',t') \tag{12}$$

This equation is valid for colloidal suspensions of particles with axially symmetric potentials, where the memory function $M(k,\Omega,\Omega',t)$ is related to the friction Eq. (6), which does not require such an approximation on L [52]. The solution of this equation for the transversal component of the van Hove function $C(k,\Omega,t)$ leads to the complex susceptibility function in the frequency domain, and it is given by $\chi(k,\omega) = (4\pi\rho^*\mu^{*2}/3)\left[S^{11}_{,0}(x) - i\omega C^{11}_{,1}(x,\omega)\right] + 1$. This equation reproduces

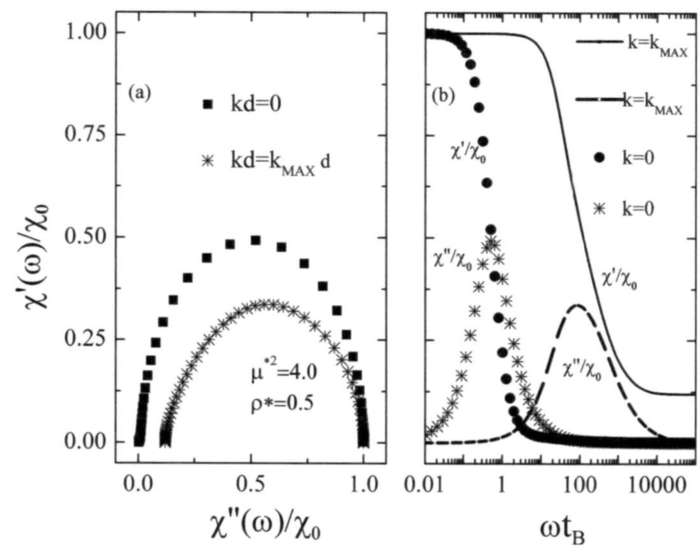

Figure 6.
Real χ' and imaginary χ'' magnetic susceptibilities of a ferrofluid normalized to the initial value of susceptibility χ_0. Picture (a) is a magnetic Cole-Cole plot and (b) depicts the susceptibility moduli versus frequency ω. The susceptibilities were evaluated at k_{max} the value of the wave number at the highest first peak of the structure factor, and $k = 0$.

similar results of a model molecular liquid of Wei and Patey [53] who, however, used a Kerr dynamical equation for *C*. Since $\chi = \chi' + i\chi''$ in **Figure 6**, we provide typical plots of the real and imaginary components of the magnetic susceptibility of the ferrofluid Fe_2O_3.

Alternative kinetics methods for low-density ferrofluids are given in [40].

3. Conclusion

This review highlights several active research lines on the structure and dynamics in ferrofluids. These investigations are motivated by many experimental observations that colloidal ferromagnetic particles both in bulk or in thin film aggregate, building several macroscopic structures of practical and scientific interest. As for all applications where colloidal materials are being processed, it is necessary to understand the interparticle interactions to have better control of the resulting structures and their properties. The interactions are known experimentally under specific conditions through the measured phase diagrams [11]; it has been found [29] that a useful potential that represents well the structure factor of ferrofluids is a Lennard-Jones plus dipolar interaction between pairs of particles. Therefore, our use of this potential in an evolutionary genetic algorithm leads to **Figure 1**, which shows transient metastable crystalline structures of a monodisperse ferrofluid at $T = 0K$. On the other hand, **Figure 2** shows the predicted structure factors using material parameters of typical Fe_2O_3 ferrofluid. There we observe that for low particle density and dipole moment, the particles are dispersed in the fluid, whereas for the highest dipole moment system, they form chains. As a consequence, there are noticeable effects on the collective diffusion coefficient as it is shown in **Figure 3,** where we notice that as the strength of the dipole moment is increased for low-density suspensions, this property drops because of particles' association in larger structures (chains). However, for higher-density suspensions, particles re-disperse again, therefore, increasing their collective diffusion. In Section 2.4, we presented the stochastic Langevin approach to describe the diffusion of a tracer particle in colloidal magnetic fluids. **Figure 4** provides theoretical predictions of this theory for the longtime translational and rotational self-diffusion coefficients Eq. (7) of a ferroparticle for the same material parameters of Fe_2O_3. We notice that the theory prediction has an agreement with Langevin dynamics simulations. In Section 2.5, we described our extension of a hydrodynamic theory of colloid dynamics [52] to include the rotational Brownian motions of ferrofluid's particles. As a result, we derived an equation of the complex susceptibility $\chi(k, w)$ that depends on the diffusion constants $D(\omega)$, $D_R(\omega)$ of a tracer particle in the colloid, which were given in Section 2.4. Thus, in **Figure 6,** we show the predicted magnetic Cole-Cole plot and real and imaginary components of the susceptibility moduli for Fe_2O_3 ferrofluid. There remain further studies to find the successful comparison of this method with well-characterized magnetic colloidal fluids and computer simulations in model ferrofluid systems.

Acknowledgements

The authors acknowledge the General Coordination of Information and Communications Technologies (CGSTIC) at CINVESTAV for providing HPC resources on the Hybrid Supercomputer "Xiuhcoatl" that has contributed to the research results reported within this chapter.

Conflict of interest

There are no conflicts to declare.

Author details

Ricardo Peredo Ortíz[1], Martin Hernández Contreras[1*] and Raquel Hernández Gómez[2]

1 Physics Department, Center for Research and Advanced Studies of the National Polytechnic Institute, CD Mexico, Mexico

2 Computation Department, Center for Research and Advanced Studies of the National Polytechnic Institute, CD Mexico, Mexico

*Address all correspondence to: marther@fis.cinvestav.mx

IntechOpen

© 2018 The Author(s). Licensee IntechOpen. This chapter is distributed under the terms of the Creative Commons Attribution License (http://creativecommons.org/licenses/by/3.0), which permits unrestricted use, distribution, and reproduction in any medium, provided the original work is properly cited.

References

[1] Rosensweig RE. Ferrohydrodynamics. Cambridge: Cambridge University Press; 1985

[2] Jordan A, Scholz R, Maier-Hauff K, Johannsen M, Wust P, Nadobny J, et al. Presentation of a new magnetic field therapy system for the treatment of human solid tumors with magnetic fluid hyperthermia. Journal of Magnetism and Magnetic Materials. 2001;**225**:118-126

[3] Kim SH, Lee SY, Yang SM, Yi GR. Self-assembled colloidal structures for photonics. NPG Asia Materials. 2011;**3**: 25-33

[4] He L, Wang H, Ge J, Yin Y. Magnetic assembly route to colloidal responsive photonic nanostructures. Accounts of Chemical Research. 2012;**45**:1431-1440

[5] Andelman D, Rosensweig RE. Modulated Phases: Review and Recent Results. The Journal of Physical Chemistry. B. 2009;**113**:3785-3798

[6] Klapp SHL. Collective dynamics of dipolar and multipolar colloids: from passive to active systems. Current Opinion in Colloid & Interface Science. 2016;**21**:76-85

[7] Tierno P. Recent advances in anisotropic magnetic colloids: Realization, assembly and applications. Physical Chemistry Chemical Physics. 2014;**16**:23515-23528

[8] Schilling R, Scheidsteger T. Mode coupling approach to the ideal glass transition of molecular liquids: Linear molecules. Physical Review E. 1997;**56**: 2932-2949

[9] Snezhko A. Non-equilibrium magnetic colloidal dispersions at liquid–air interfaces: dynamic patterns, magnetic order and self-assembled swimmers. Journal of Physics: Condensed Matter. 2011;**15**:153101 (1-21)

[10] Ramírez-González PE, Medina Noyola M. Glass transition in soft-sphere dispersions. Journal of Physics: Condensed Matter. 2009;**21**:075101 (1-13)

[11] Yethiraj A, van Blaaderen A. A colloidal model system with an interaction tunable from hard sphere to soft and dipolar. Nature. 2003;**421**: 513-517

[12] Klokkenburg M, Erne BH, Meeldijk JD, Wiedenmann A, Pethukov AV, Dullens RPA, et al. In situ imaging of field-induced hexagonal columns in magnetite ferrofluids. Physical Review 36 Letters. 2006;**97**:185702 (1-4)

[13] Cousin F, Dubois E, Cabuil V. Tuning the interactions of a magnetic colloid suspension. Physical Review E. 2003;**68**:021405 (1-9)

[14] Pham AT, Zhuang Y, Detwiler P, Socolar JES, Charbonneau P, Yellen BB. Phase diagram and aggregation dynamics of a monolayer of paramagnetic colloids. Physical Review E. 2017;**95**:052607 (1-12)

[15] Hynninen AP, Dijkstra M. Phase Diagram of dipolar hard and soft spheres: Manipulation of colloidal crystal structures by an external field. Physical Review Letters. 2005;**94**: 138303 (1-4)

[16] Spiteri L, Messina R. Dipolar Crystals: The crucial role of the clinohexagonal prism phase. Physical Review Letters. 2017;**119**:155501 (1-4)

[17] Weis JJ, Levesque D. Advanced Computer Simulation Approaches for Soft Matter Sciences II. Berlin Heidelberg: Springer; 2005. pp. 163-225

[18] Wagner J, Fischer B, Autenrieth T. Field induced anisotropy of charged magnetic colloids: A rescaled mean spherical approximation study.

The Journal of Chemical Physics. 2006; **124**:114901 (1-10)

[19] Wagner J, Fischer B, Autenrieth T, Hempelmann R. Structure and dynamics of charged magnetic colloids. Journal of Physics: Condensed Matter. 2006;**18**:S2697-S2711

[20] Martin GAR, Bradbury A, Chantrell RW. A comparison between a three-dimensional Monte Carlo model and a hypernnetted chain approximation of a ferrofluid. Journal of Magnetism and Magnetic Materials. 1987;**65**:177-180

[21] Bradbury A, Martin GAR, Chantrell RW. Zero field particle correlations in a ferrofluid using Monte Carlo simulations and the hypernetted chain approximation. Journal of Magnetism and Magnetic Materials. 1987;**69**:5-11

[22] Hayter JB, Pynn R. Structure factor of a magnetically saturated ferrofluid. Physical Review Letters. 1982;**49**:1103-1106

[23] Pynn R, Hayter JB, Charles SW. Determination of ferrofluid structure by neutron polarization analysis. Physical Review Letters. 1983;**51**:710-713

[24] Bacri JC, Cebers A, Bourdon A, Demouchy G, Heegaard BM, Perzynski R. Forced rayleigh experiment in a magnetic fluid. Physical Review Letters. 1995;**74**:5032-5935

[25] Gazeau F, Dubois E, Bacri JC, Boue F, Cebers A, Perzynski R. Anisotropy of the structure factor of magnetic fluids under a field probed by small-angle neutron scattering. Physical Review E. 2002;**65**:031403 (1-15)

[26] Bacri JC, Cebers A, Bourdon A, Demouchy G, Heegaard BM, Kashevsky B, et al. Transient grating in a ferrofluid under magnetic field: Effect of magnetic interactions on the diffusion coefficient of translation. Physical Review E. 1995; **52**:3936-3942

[27] Meriguet G, Dubois E, Bourdon A, Demouchy G, Dupuis V, Perzynski R. Forced Rayleigh scattering experiments in concentrated magnetic fluids: effect of interparticle interactions on the diffusion coefficient. Journal of Magnetism and Magnetic Materials. 2005;**289**:39-42

[28] Lal J, Abernathy D, Auvray L, Diat O, Grubel G. Dynamics and correlations in magnetic colloidal systems studied by X-ray photon correlation spectroscopy. European Physical Journal E: Soft Matter and Biological. Physics. 2001;**4**: 263-271

[29] Meriguet G, Dubois E, Jardat M, Bourdon A, Demouchy G, Dupuis V, et al. Understanding the structure and the dynamics of magnetic fluids: coupling of experiment and simulation. Journal of Physics: Condensed Matter. 2006;**18**:S2685-S2696

[30] Meriguet G, Jardat M, Turq P. Structural properties of charge-stabilized ferrofluids under a magnetic field: A Brownian dynamics study. The Journal of Chemical Physics. 2004;**121**: 6078-6085

[31] Meriguet G, Jardat M, Turq P. Brownian dynamics investigation of magnetization and birefringence relaxations in ferrofluids. The Journal of Chemical Physics. 2005;**123**:144915 (1-8)

[32] Hernández-Contreras M, Ruíz Estrada H. Transport properties of ferrofluids. Physical Review E. 2003;**68**: 031202 (1-7)

[33] Peredo-Ortíz R, Hernández Contreras M, Hernández Gómez R. Diffusion in monodisperse ferrofluids. Journal of Physics: Conference Series. 2017;**792**:012099 (1-11)

[34] Peredo-Ortíz R, Hernández Contreras M. Diffusion microrheology of ferrofluids. Revista Mexicana de Fisica. 2018;**64**:82-93

[35] Mertelj A, Rešetič A, Gyergyek S, Markovec D, Čopič M. Anisotropic microrheological properties of chain-forming magnetic fluids. Soft Matter. 2011;7:125-131

[36] Yendeti B, Thirupathi G, Vudaygiri A, Singh R. Field-dependent anisotropic microrheological and microstructural properties of dilute ferrofluids. European Physical Journal E: Soft Matter and Biological Physics. 2014;**37**: 10 (1-8)

[37] Mazoyer S, Ebert F, Maret G, Keim P. Dynamics of particles and cages in an experimental 2D glass former. Europhysics Letters. 2009;**88**:66004 (1-6)

[38] Lacoste D, Lubensky TC. Phase transitions in a ferrofluid at magnetic-field-induced microphase separation. Physical Review E. 2001;**64**:041506 (1-8)

[39] de Gennes PG, Pincus PA. Pair correlations in a ferromagnetic colloid. Physics Kondens Matter. 1970;**11**:189-194

[40] Odenbach S. Colloidal Magnetic Fluids. 1st ed. Berlin: Springer; 2009

[41] Tlusty T, Safran SA. Defect-induced phase separation in dipolar fluids. Science. 2000;**290**:1328-1331

[42] Sreekumari A, Ilg P. Slow relaxation in structure-forming ferrofluids. Physical Review E. 2013;**88**:042315 (1-9)

[43] Russel WB, Saville DA, Showalter WR. Colloidal Dispersions. New York NY: Cambridge University Press; 1989

[44] Israelachvili JN. van der Waals dispersion force contribution to works of adhesion and contact angles on the basis of macroscopic theory. Journal of the Chemical Society, Faraday Transactions 2. 1973;**69**:1729-1738

[45] Fornleitner J, Loverso F, Kahl G, Likos CN. Genetic algorithms predict formation of exotic ordered configurations for two-component dipolar monolayers. Soft Matter. 2008; **4**:480-484

[46] Autenrieth T, Robert A, Wagner J, Grubel G. The dynamic behavior of magnetic colloids in suspension. Journal of Applied Crystallography. 2007;**40**: 5250-5253

[47] Pyanzina E, Kantorovich S, Cerda JJ, Ivanov A, Holm C. How to analyze the structure factor in ferrofluids with strong magnetic interactions: a combined analytic and simulation approach. Molecular Physics. 2009;**107**: 571-590

[48] Kirchoff T, Lowen H, Klein R. Dynamical correlations in suspensions of charged rodlike macromolecules. Physical Review E. 1996;**53**:5011-5022

[49] Mason TG, Weitz DA. Optical measurements of frequency-dependent viscoelastic moduli of complex fluids. Physical Review Letters. 1995;**74**:1250-1253

[50] Mazur P, Geigenmüller U. A simple formula for the short-time self-diffusion coefficient in concentrated suspensions. Physica A. 1987;**146**:657-661

[51] Roeben E, Roeder L, Teusch S, Effertz M, Deiters UK, Schmidt AM. Magnetic particle nanorheology. Colloid & Polymer Science. 2014;**292**:2013-2023

[52] Hess W, Klein R. Generalized hydrodynamics of systems of Brownian particles. Advances in Physics. 1983;**32**: 172-283

[53] Wei D, Patey GN. Rotational motion in molecular liquids. The Journal of Chemical Physics. 1989;**91**:7113-7129

Chapter 4

The Study of Magneto-Convection Heat Transfer in a Partially Open Cavity Based on LBM

Raoudha Chaabane

Abstract

A free convection heat transfer in a sinusoidally heated enclosure filled with conducting fluid is presented in this chapter by Lattice Boltzmann Method (LBM). The horizontal walls in the enclosures are insulated and there is an opening part on the right wall. The right non-open parts of the vertical wall of the square cavity are maintained at constant cold temperature and the left wall of the cavity is sinusoidally heated. The cavity is get under a uniform in-plane magnetic field. The main aim of this study is to highlight the effectiveness of the LBM mesoscopic approach to predict the effects of pertinent parameters such as the Hartmann number varying from 0 to 150 where Rayleigh number is fixed at moderate value of 10^5 on flow patterns. This in-house numerical code used in this chapter is ascertained and a good agreement with literature is highlighted. The appropriate validation with previous numerical investigations demonstrated that this attitude is a suitable method and a powerful approach for engineering MHD problems. Findings and results show the alterations of Hartmann number that influence the isotherms and the streamlines widely.

Keywords: sinusoidal thermal boundary condition, MHD, partially open cavity, free convection, LBM

1. Introduction

Convective flow and heat transfer in an open cavity has been studied due to the extensive range of applications in engineering science and technology that consider various combinations of imposed temperature gradients and enclosure sketches [1–10]. Open cavity with a modified linear or sinusoidal thermal boundary condition is encountered in many practical engineering and industrial applications, such as solar energy collection, cooling of electronic devices, material processing, grain storage, flow and heat transfer in solar ponds, high-performance insulation for buildings, dynamics of lakes, reservoirs and cooling ponds, crystal growing, float glass production, metal casting, food processing, galvanizing, metal coating, and so on.

Besides, pertinent useful numerical research works had been conducted to simulate the MHD free convection under nonuniform thermic boundaries where recent attention has been intensively focused on the cases of mixed boundary conditions [11–14]. However, few results have been reported for free convection caused

IntechOpen

simultaneously by both external magnetic field in partially open enclosures subjected to a sinusoidal temperature variation in the left vertical wall although problems of this type are frequently important, and their study is necessary for understanding the performance of complex magneto-convection flow and heat transfer.

MHD forces generated from the interaction of induced electric currents with an applied external magnetic field can alter the flow of an electrically conducting fluid in the presence of magnetic field. An externally imposed magnetic field is an important tool used to control melting flows that grow bulk crystal in semiconductor's applications. A main purpose of electromagnetic control is to stabilize the flow and suppress oscillatory instabilities, which degrades the resulting crystal. In literature, wide ranges of investigations were investigated by researchers in MHD free convection [15–30]. In such complex geometry, the balance is achieved by inertial, viscous, electromagnetic, and buoyancy forces; finding a numerical efficient tool to predict flow and heat pattern inside MHD cavities is a crucial aim for industrial related engineering applications.

The effect of Ra number on free convection MHD in an open cavity was investigated in [29]. Nonetheless several investigations in MHD free convection inside partially open enclosure with sinusoidally wall have been carried out yet.

The mesoscopic approach called Boltzmann method (LBM) joined the microscopic models and the macroscopic dynamics of a fluid. It can recover the Navier-Stokes equation by using the Chapman-Enskog expansion [39]. LB method is not very demanding in terms of memory requirement. In addition, in terms of computational speed, the algorithm is generally simpler and therefore faster than many traditional CFD schemes. It is easy for parallel computation and for implementation of irregular boundary conditions, which is a sought task in many-needed engineering geometry.

To our best knowledge, no previous study on effects of sinusoidally heated boundary on free convection in a partially open MHD enclosure cavity with the LBM had already been studied so far. The main aim of this chapter is to study the effects of linearly heated wall on flow field and temperature distribution in an open MHD cavity filled.

The main aim of the present study is to demonstrate the use of the Lattice Boltzmann Method (LBM) [31–40] for MHD with a simple and clear statement and also solve MHD free convection as a left sinusoidally heated side mixed with a partially open right wall filled with a conducting fluid. Hartmann number varies in a wide range from 0 to 150. First, the results of LBM are validated with previous numerical investigations. Effects of Rayleigh, Hartmann number, and various positions of the open side on flow field and temperature distribution are considered simultaneously.

The proposed configuration with sinusoidal temperatures on the left side wall of a partially open cavity in the presence of a magnetic field has not been focused. A major objective of the present study is to examine the magneto-convection in this configuration filled with a conducting fluid confined between two horizontal walls, which are thermally insulated. The effect of the open side on fluid flow and heat transfer is studied numerically.

2. Governing equations and mathematical formulation

2.1 Problem statement

The considered geometries of the present problem are shown in **Figure 1**. They display a two-dimensional closed, open or partially open east side cavity with the height of H. A constant, linear, or sinusoidal temperature is imposed along the left vertical wall. Then opening side boundaries are correlated with temperature

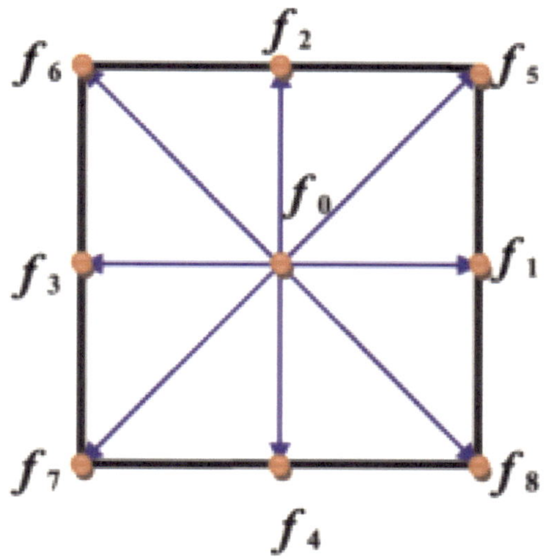

Figure 1.
The standard D_2Q_9 LBM lattice.

conducting fluid at the cold temperature (Tc). The north and south horizontal boundaries are adiabatic. We consider a horizontal uniform magnetic field applied to a two-dimensional Newtonian, laminar, and incompressible conducting fluid. The radiation effects, the viscous dissipation, and Joule heating are neglected in the present study. The thermophysical properties of the conducting fluid are constant, and the density variation in the liquid gallium is approximated by the standard Boussinesq (**Figure 2**).

2.2 Governing equations

Governing equations for MHD free convection are written in terms of the macroscopic variable depending on position x,y as:

Continuity equation

$$\frac{\partial u}{\partial x}+\frac{\partial v}{\partial y}=0 \tag{1}$$

Momentum equations

$$\rho\left(u\frac{\partial u}{\partial x}+v\frac{\partial u}{\partial y}\right)=-\frac{\partial p}{\partial x}+\mu\left(\frac{\partial^2 u}{\partial x^2}+\frac{\partial^2 u}{\partial y^2}\right)+F_x \tag{2}$$

$$\rho\left(u\frac{\partial v}{\partial x}+v\frac{\partial v}{\partial y}\right)=-\frac{\partial p}{\partial y}+\mu\left(\frac{\partial^2 v}{\partial x^2}+\frac{\partial^2 v}{\partial y^2}\right)+F_y \tag{3}$$

Energy equation

$$u\frac{\partial T}{\partial x}+v\frac{\partial T}{\partial y}=\alpha\left(\frac{\partial^2 T}{\partial x^2}+\frac{\partial^2 T}{\partial y^2}\right) \tag{4}$$

where $\nu=\mu/\rho$ is the kinematic viscosity and F_x and F_y are the body forces at horizontal and vertical directions, respectively, and they are defined as follows [30]:

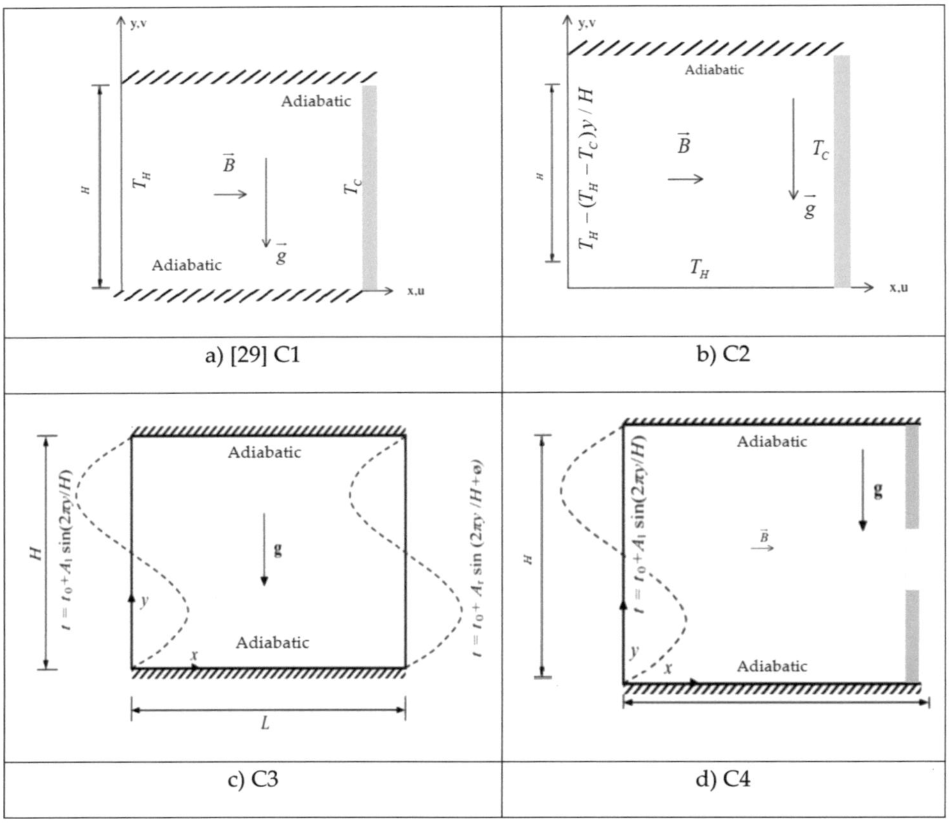

Figure 2.
Used configurations C1–4 with different boundary conditions.

$$F_x = R\left(v \sin\gamma \cos\gamma - u \sin^2\gamma\right) \tag{5}$$

$$F_y = R\left(u \sin\gamma \cos\gamma - v \cos^2\gamma\right) + \rho g \beta (T - T_m) \tag{6}$$

where the Ha number is defined as:

$$Ha = H B_x \sqrt{\sigma/\mu} \tag{7}$$

The LBM method [31–40] with standard, two-dimensional, nine velocities (D2Q9) for flow and temperature is used in this chapter (**Figure 1a**); for completeness, only a brief discussion is given in the following paragraphs. The Bhatnagar-Gross-Krook (BGK) approximation Lattice Boltzmann equation with external forces F_{tot} can be written as:

$$f_\alpha(\mathrm{r} + \mathrm{e}_\alpha \Delta t, t + \Delta t) = f_\alpha(\mathrm{r}, t) - \frac{\Delta t}{\tau_f}\left[f_\alpha(\mathrm{r}, t) - f_\alpha^{eq}(\mathrm{r}, t)\right] + \Delta t\, \mathrm{F}_{tot} \tag{8}$$

where $f_\alpha(\mathrm{r}, t)$ is the particle distribution defined for the finite set of the discrete particle velocity vectors e_α. r and t are the coordinates of Eulerian node and time.

where τ_f is the relaxation time and $f_\alpha^{eq}(\mathrm{r}, t)$ is the local equilibrium distribution function.

The equilibrium distribution can be formulated as:

$$f_\alpha^{eq} = w_\alpha \rho \left(1 + \frac{\mathrm{e}_\alpha.\mathrm{u}}{c_s^2} + \frac{(\mathrm{e}_\alpha.\mathrm{u})^2}{2c_s^4} - \frac{\mathrm{u.u}}{2c_s^2}\right) \tag{9}$$

where u and ρ are the macroscopic velocity and density, respectively, and w_α are the values of the weighting constant factors for e_α that must be assigned as $w_0 = 4/9$ (rest-particle) for $|e_0| = 0$, $w_{1-4} = 1/9$ for $|e_{1-4}| = 1$, and $w_{5-9} = 1/36$ for $|e_{5-9}| = \sqrt{2}$.

where $c = \Delta x/\Delta t$, Δx and Δt are the lattice space and the lattice time step size, respectively, which are set to unity.

For scalar function (temperature), another distribution is defined:

$$g_\alpha(\mathrm{r} + \mathrm{e}_\alpha \Delta t, t + \Delta t) = g_\alpha(\mathrm{r}, t) - \frac{\Delta t}{\tau_g}\left[g_\varepsilon(\mathrm{r}, t) - g_\alpha{}^{eq}(\mathrm{r}, t)\right] \tag{10}$$

The equilibrium distribution function can be written as:

$$g_\alpha^{eq} = w_\alpha T\left(1 + \frac{\mathrm{e}_\alpha.\mathrm{u}}{c_s^2} + \frac{(\mathrm{e}_\alpha.\mathrm{u})^2}{2c_s^4} - \frac{\mathrm{u.u}}{2c_s^2}\right) \tag{11}$$

For momentum equation, we have:

$$\tau_f = 0.5 + \left(3\nu/c^2\Delta t\right) \tag{12}$$

where ν is the kinematic viscosity and for the scalar:

$$\tau_g = 0.5 + \left(3\alpha/c^2\Delta t\right) \tag{13}$$

where α is the diffusion coefficient (thermal diffusion coefficient) and Δt is the lattice time step. The buoyancy force term is added as an extra source term to Eq. (1) as:

$$F_\alpha = 3w_\alpha g_y \beta \Delta T \tag{14}$$

where g_y, β, and ΔT are gravitational acceleration, thermal expansion coefficient, and temperature difference, respectively.

In the LBM, the total force is

$$F = F_x + F_y \tag{15}$$

$$F_x = 3w_k\rho\left[R(v \sin\gamma \cos\gamma) - u \sin^2\gamma)\right] \tag{16}$$

$$F_y = 3w_k\rho\left[g_y\beta(T - T_m) + R(u \sin\gamma \cos\gamma) - v \cos^2\gamma\right] \tag{17}$$

where $R = \mu Ha^2$ and γ is the direction of the magnetic field.

After completing streaming and collision processes where the Boussinesq approximation is considered for free convection, the macroscopic fluid quantities, namely, the macro density, velocity (obtained through moment summations in the velocity space), and temperature are computed.

$$\rho(\mathrm{r}, t) = \sum_k f_k(\mathrm{r}, t) \tag{18}$$

$$\mathrm{u}(\mathrm{r}, t) = \sum_k \mathrm{e}_k f_k(\mathrm{r}, t)/\rho(\mathrm{r}, t) \tag{19}$$

$$T = \sum_k g_k(\mathrm{r}, t) \tag{20}$$

3. Results and discussion

In all cases, both the computations for flow and temperature fields are based on the D2Q9 LBM approach.

We have validated our in-house Fortran computer code for the free convection in a square open cavity with insulated horizontal walls filled with air with a uniform right (west) vertical temperature by reference [29]. The obtained numerical results are compared with the numerical ones reported in [29]. It can be seen from **Figure 3** that there is a good agreement for the distribution of streamlines and isotherms among the present solution and literature and this for Pr = 0.71, Ha = 0, and Ra = 10^5. The used configuration is C1 (**Figure 2**).

In **Figure 4**, the considered matter is MHD free convection in an open cavity with linearly heated west wall (C2), which is filled with liquid gallium (Pr = 0.025) and Ha = 150. As shown in **Figure 4**, an increase in Rayleigh number makes the thermal boundary layer to become narrower for the constant parameters Pr = 0.025 and Ha = 150.

Figure 5 displays steady-state contour maps for the isotherms and the streamline contours at Pr = 0.025 for a high Rayleigh number Ra = 10^6 and high Hartmann

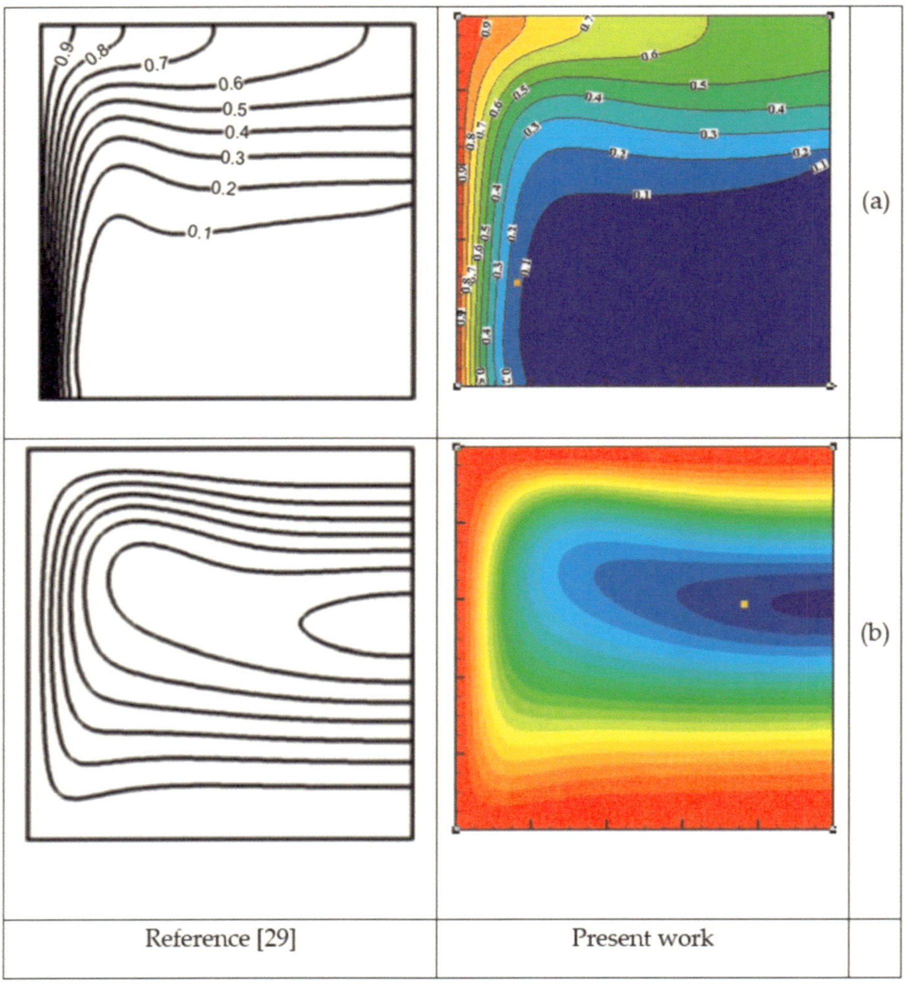

Figure 3.
Comparison of the steady state isotherms (a) and streamlines (b) at Pr = 0.71 for Ha = 0 and Ra = 10^5 between [29] and the present work.

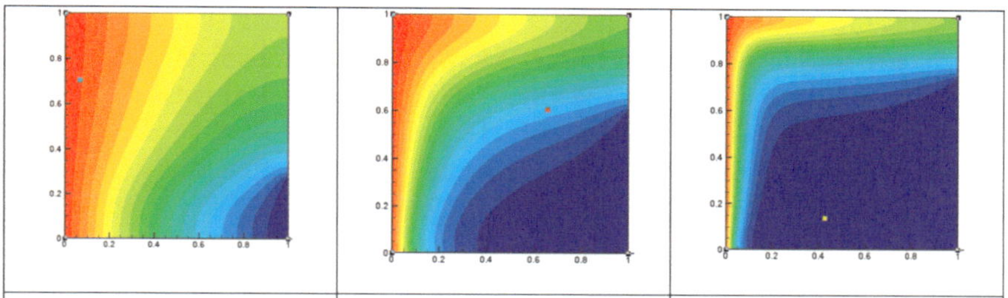

$Ra=10^5$	$Ra=10^6$	$Ra=10^7$

Figure 4.
Steady-state isotherms (b) at Pr = 0.025 and Ha = 150 for different Rayleigh numbers.

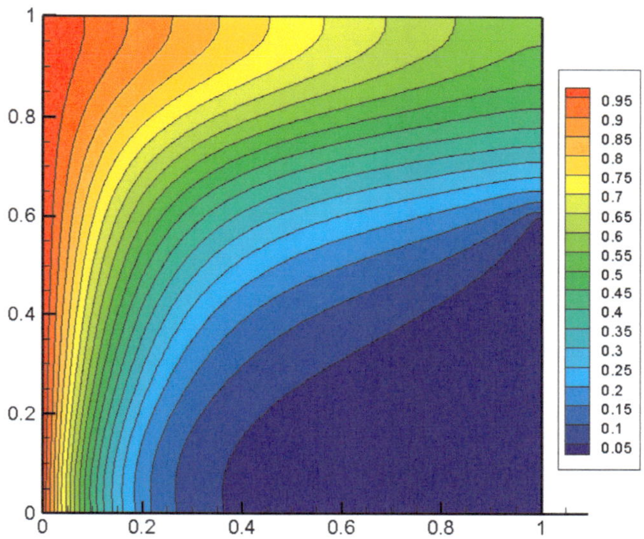

(a)

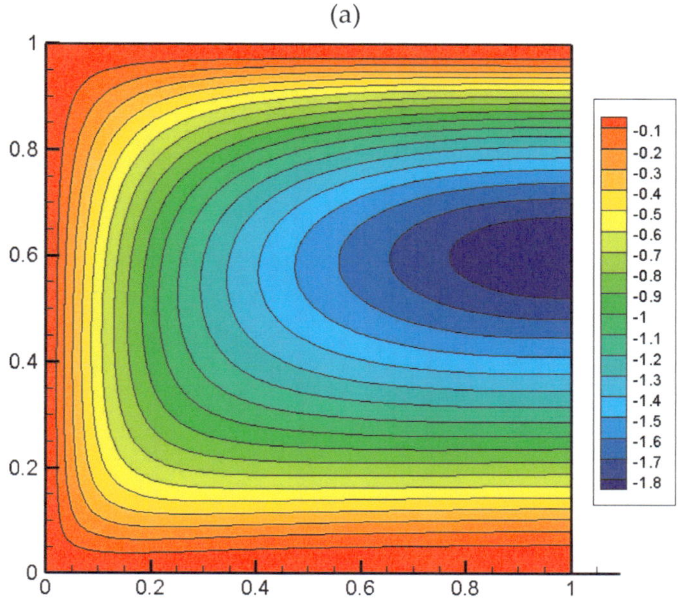

(b)

Figure 5.
Steady-state isotherms (a) and isotherms (b) at Pr = 0.025, Ra = 106, and Ha = 150.

number Ha = 150. Configuration C2 is considered for this case. The effect of the open right side wall is depicted in **Figure 5a** via streamline display.

In this chapter, we aim to test the ability of the LBM to deal with a future complex configuration (C4), which is a MHD partially open cavity with sinusoidal input excitation on the west wall. The horizontal walls are adiabatic. The cavity is

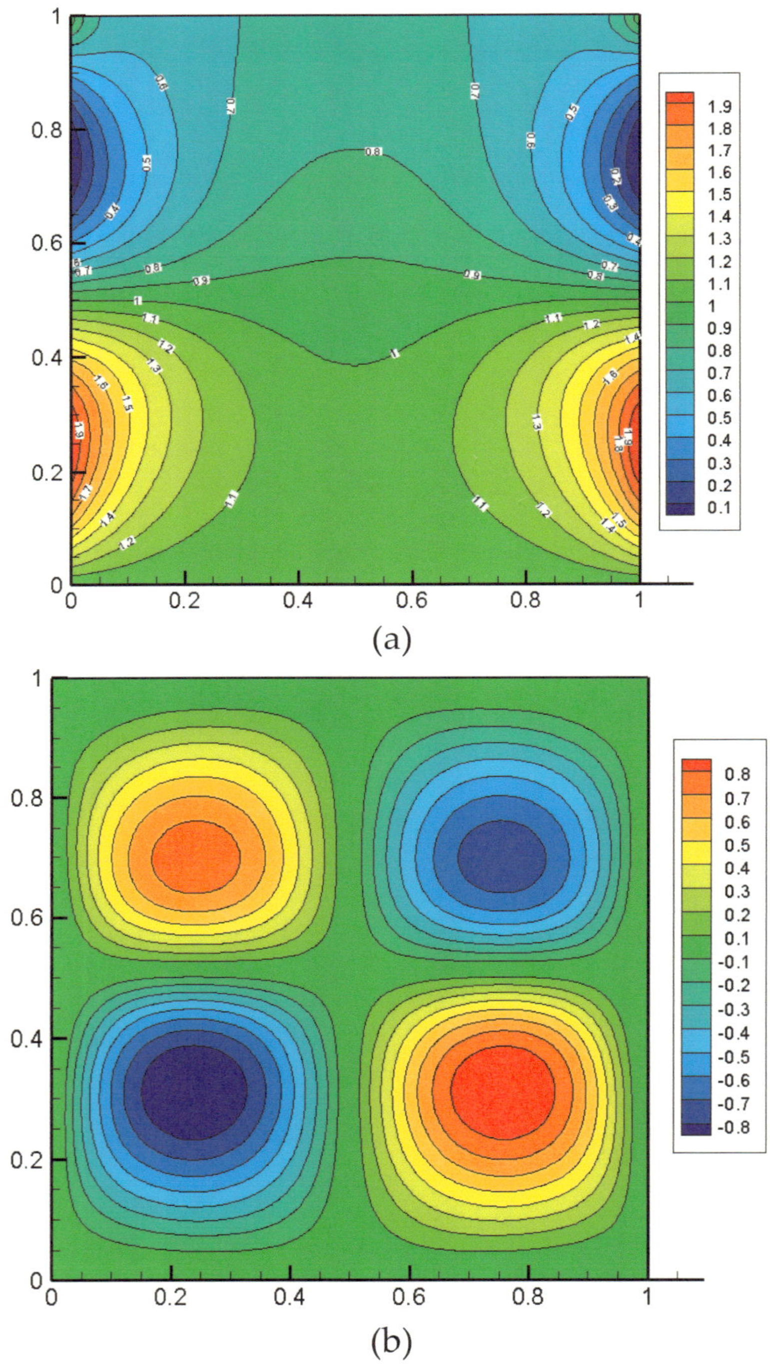

Figure 6.
Steady-state isotherms (a) and streamlines (b) of sinusoidally heated side walls in a partially open MHD cavity for Ha = 150, Pr = 0.025 and Ra = 10^5.

filled with liquid gallium, and the simulation is done for Ha = 50 and a moderate Rayleigh number of Ra = 10^5.

Many researchers were considering a sinusoidal heating wall in their simulation [11–14]. For brevity, we resume their important findings.

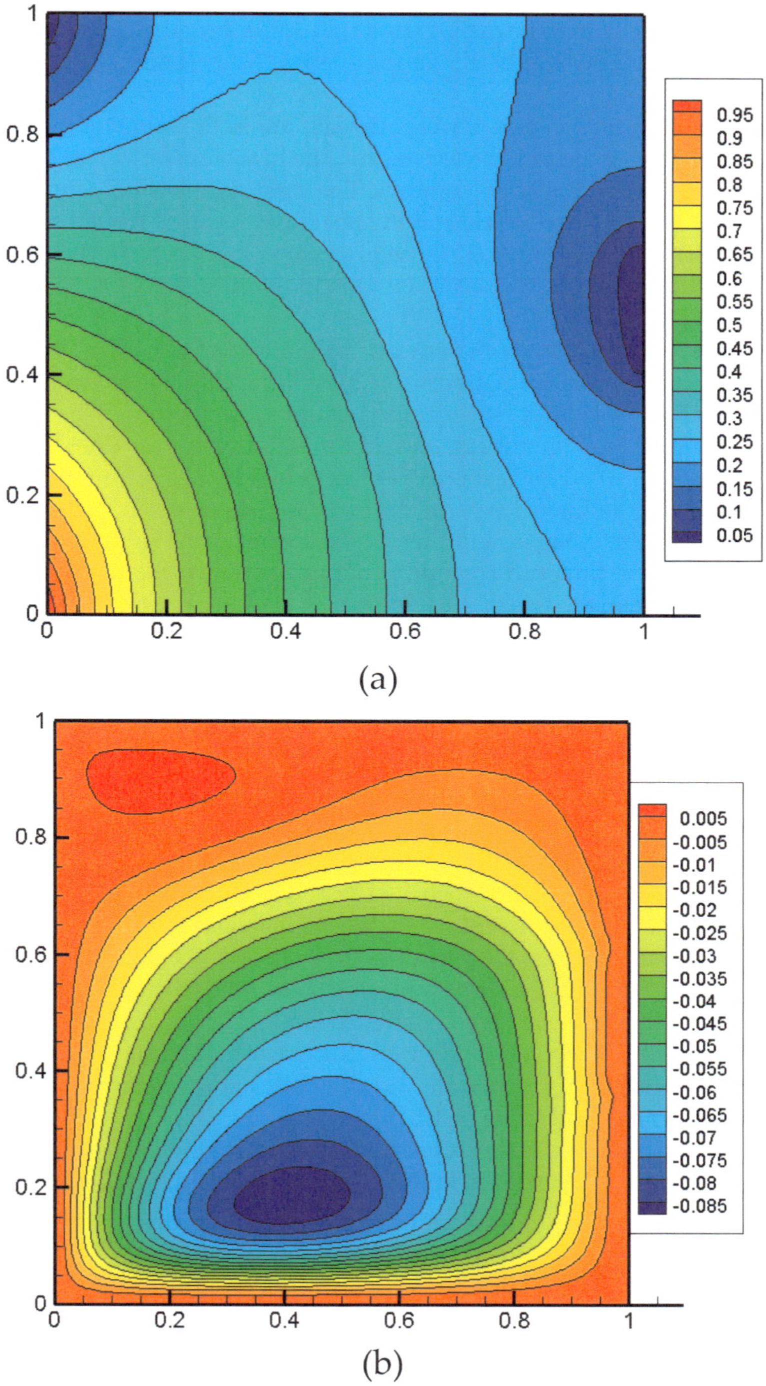

Figure 7.
Steady-state isotherms (a) and streamlines (b) of sinusoidally heated right side wall partially open MHD cavity for Ha = 150, Pr = 0.025 and Ra = 10^6.

In literature, we find that the heat transfer rate was increased as the amplitude ratio of the sinusoidal excitation increases with Rayleigh and Hartmann numbers.

For a uniform heating wall, the heat transfer rate remains low, which makes that the nonuniform heating of both walls is advised for enhancing and improving heat transfer. It is proven that the heat transfer rate is increased first and then decreased on increasing the phase deviation from 0 to pi [11–14].

In addition, with a phase deviation from 0 to 3pi/4, heat transfer rate is enhanced for all Rayleigh numbers, and the average Nusselt number reaches its highest value at 3pi/4.

A further finding proves that when both walls are in the same temperature distribution in the absence of phase deviation, the heat transfer rate is low for all values of Hartmann and Rayleigh numbers. The right wall is widely influenced by the variation of the amplitude ratio and the phase deviation of the sinusoidal temperature distribution. However, those physical variations have very little effect on the left wall. Besides, an increase in Hartmann number decreases the heat transfer [11–14].

In the present chapter, we seek to deal with a new configuration C4. For this goal, we deal first with the configuration (C3) of **Figure 2**. The sidewalls of the cavity have spatially varying sinusoidal temperature distributions. The horizontal walls are adiabatic. Simulation is established for MHD cavity for Ha = 150, Pr = 0.025, and Ra = 10^6. The heat transfer rate is highlighted within the evolution of isotherms and streamlines inside the cavity in **Figure 6**.

Figure 7 shows the dynamic and thermic behavior of configuration C4. Numerical results in terms of flow and thermic structure show that the flow within the cavity takes place owing to the thermic buoyancy effects caused by the sinusoidally heated right wall. In C3, the flow is characterized by a symmetric multicellular behavior in which the recirculating eddies or cells of relatively high velocity are formed within the enclosure and this in the presence of a high Hartmann Ha = 150 and a high Rayleigh number Ra = 10^5.

In C4, the presence of the partially open sidewall changes the flow and heat transfer. We note the presence of a one dominant cell in the core region of the cavity and a little cell that occurs at upper left side.

As forecasted, because of the partially open side effects, the temperature field was sketched by a noticeable drop in its behavior near the open side of the cavity and these both in flow and heat behaviors. Besides, we highlight that the temperature contour maps lose the sinusoidal behavior as they move to the partially open side.

We infer that a partially open sidewall has the tendency to control efficiently the movement of the fluid in such given configuration (C4).

We subject in the convergence criterion that the relative change in two successive iterates of the solution (temperate and velocities) at each computational point be below a prescribed small value of 10^{-6}.

For all simulations, the criterion convergence is considered to be reached, for velocity and temperature when the following convergence is satisfied:

$$\left|\xi^{\sigma+1} - \xi^{\sigma}\right| \leq 10^{-6} \tag{21}$$

where ξ is velocity or temperature and σ is the iteration number.

4. Conclusions

To the author's knowledge, studies have thus far addressed a mesoscopic approach in an MHD open cavity with sinusoidally heated wall (**Figure 2**). The objective of the present chapter is therefore to predict dynamic and thermic heat transfer in a crucial engineering application. The cavity is investigated at the high Ra number of 10^5, high Ha number (Ha = 50), and Pr number of 0.025. The present chapter extends the study to deal with free convection in MHD open cavity with sinusoidal heated west wall which is filled with liquid gallium with Ha = 50 for Ra = 10^5. Numerical study has been made of free convection in a square enclosure with spatially varying sinusoidal temperature distributions on the vertical left sidewall, whereas the horizontal walls are thermally insulated. The right wall is a partially open one. Lattice Boltzmann method is considered for flow and heat transfer simulation of this problem inside the cavity. After validation, the present in-house Fortran code is extended to deal with the present complex geometry in order to highlight the workability and the ability of LBM to deal with such mixed boundary condition sketch. This investigation demonstrated ability of LBM for simulation of different boundary conditions at various elements affecting the stream in a partially open cavity with sinusoidal heating vertical wall. An analysis of the opening mass flow is highlighted in the dynamic and thermal behavior of the streamlines and the isotherms.

Nomenclature

C	lattice speed
c_i	discrete particle speeds
c_p	specific heat at constant pressure
F	external forces
f^{eq}	equilibrium density distribution functions
g^{eq}	equilibrium internal energy distribution functions
g	gravity
G	buoyancy per unit mass
H	enclosure height
Ma	Mach number
Nu	Nusselt number
Pr	Prandtl number
Ra	Rayleigh number
T	temperature
x,y	Cartesian coordinates
X	horizontal length of the cavity
Y	vertical length of the cavity
Ha	Hartmann number

Greek letters

ω_i	weighted factor indirection i
β	thermal expansion coefficient
τ	relaxation time
ν	kinematic viscosity
Δt	time increment
α	thermal diffusivity

Subscripts

av	average
H	hot
C	cold
b	bottom

Author details

Raoudha Chaabane[1,2]

1 Laboratory of Thermal and Energetic Systems Studies (LESTE), National School of Engineering of Monastir, University of Monastir, Tunisia

2 Preparatory Institute of Engineering Studies of Monastir (IPEIM), University of Monastir, Tunisia

*Address all correspondence to: raoudhach@gmail.com

IntechOpen

© 2019 The Author(s). Licensee IntechOpen. This chapter is distributed under the terms of the Creative Commons Attribution License (http://creativecommons.org/licenses/by/3.0), which permits unrestricted use, distribution, and reproduction in any medium, provided the original work is properly cited.

References

[1] Lauriat G, Desrayaud G. Effect of surface radiation on conjugate natural convection in partially open enclosures. International Journal of Thermal Sciences. 2006;**45**(4):335-346

[2] Bilgen E, Oztop H. Natural convection heat transfer in partially open inclined square cavities. International Journal of Heat and Mass Transfer. 2005;**48**(8):1470-1479

[3] Chan YL, Tien CL. A numerical study of two-dimensional laminar natural convection in shallow open cavities. International Journal of Heat Mass Transfer. 1985;**28**(3):603-612

[4] Xia JL, Zhou ZW. Natural convection in an externally heated partially open cavity with a heated protrusion. FED-vol. 143/HTD, Vol. 232. Measurement and Modeling of Environmental Flows—ASME. 1992;**232**:201-208

[5] Angirasa D, Eggels JGM, Nieuwstadt FTM. Numerical simulation of transient natural convection from an isothermal cavity open on a side. Numerical Heat Transfer, Part A: Applications. 1995;**28**(6):755-768

[6] Mohamad AA. Natural convection in open cavities and slots. Numerical Heat Transfer, Part A: Applications. 1995;**27**:705-716

[7] Kennedy P, Zheng R. Flow Analysis of Injection Molds. Munich: Hanser; 2013

[8] Mohamad AA, El-Ganaoui M, Bennacer R. Lattice Boltzmann simulation of natural convection in an open ended cavity. International Journal of Thermal Sciences. 2009;**48**:1870-1875

[9] Mohamad AA, Bennacer R, El-Ganaoui M. Double dispersion natural convection in an open end cavity simulation via Lattice Boltzmann method. International Journal of Thermal Sciences. 2010;**49**:1944-1953

[10] Polat O, Bilgen E. Laminar natural convection in inclined open shallow cavities. International Journal of Thermal Sciences. 2002;**41**:360-368

[11] Bilgen E, Yedder RB. Natural convection in enclosure with heating and cooling by sinusoidal temperature profiles on one side. International Journal of Heat and Mass Transfer. 2007;**50**:139-150

[12] Sarris IE, Lekakis I, Vlachos NS. Natural convection in a 2D enclosure with sinusoidal upper wall temperature. Numerical Heat Transfer A. 2002;**42**:513-530

[13] Varol Y, Oztop HF, Pop I. Numerical analysis of natural convection for a porous rectangular enclosure with sinusoidally varying temperature profile on the bottom wall. International Communications in Heat and Mass Transfer. 2008;**35**:56-64

[14] Saeid NH, Yaacob Y. Natural convection in a square cavity with spatial side wall temperature variation. Numerical Heat Transfer A. 2006;**49**:683-697

[15] Fidaros D, Grecos A, Vlachos N. Development of numerical tool for 3D MHD natural convection. Annex xx. pp. 73-74

[16] Sathiyamoorthy M, Chamkha A. Effect of magnetic field on natural convection flow in a liquid gallium filled square cavity for linearly heated side wall(s). International Journal of Thermal Sciences. 2010;**49**:1856-1865

[17] Alchaar S, Vasseur P, Bilgen E. Natural convection heat transfer in a rectangular enclosure with a transverse

magnetic field. Journal of Heat Transfer. 1995;**117**:668-673

[18] Garandet J, Alboussiere T, Moreau R. Buoyancy drive convection in a rectangular enclosure with a transverse magnetic field. International Journal of Heat and Mass Transfer. 1992;**35**: 741-748

[19] Rudraiah N, Barron R, Venkatachalappa M, Subbaraya C. Effect of a magnetic field on free convection in a rectangular enclosure. International Journal of Engineering Science. 1995;**33**:1075-1084

[20] Cowley M. Natural convection in rectangular enclosures of arbitrary orientation with magnetic field vertical. Magnetohydrodynamics. 1996;**32**: 390-398

[21] Teamah M. Hydro-magnetic double-diffusive natural convection in a rectangular enclosure with imposing an inner heat source or sink. Alexandria Engineering Journal. 2006;**45**(4): 401-415

[22] Ozoe H, Okada K. The effect of the direction of the external magnetic field on the three-dimensional natural convection in a cubic enclosure. International Journal of Heat and Mass Transfer. 1989;**32**:1939-1953

[23] Ece M, Buyuk E. Natural convection flow under a magnetic field in an inclined rectangular enclosure heated and cooled on adjacent walls. Fluid Dynamics Research. 2006;**38**(5): 546-590

[24] Al-Najem N, Khanafer K, El-Refaee M. Numerical study of laminar natural convection in tilted enclosure with transverse magnetic field. International Journal of Numerical Methods for Heat and Fluid Flow. 1998;**8**:651-672

[25] Jalil J, Al-Taey K. MHD turbulent natural convection in a liquid metal filed square enclosure. Emirates Journal for Engineering Research. 2007;**12**(2):31-40

[26] Gelfgat A, Bar-Yoseph P. The effect of an external magnetic field on oscillatory instability of convective flows in a rectangular cavity. Physics of Fluids. 2001;**13**(8):2269-2278

[27] Aleksandrova S, Molokov S. Three-dimensional buoyant convection in a rectangular cavity with differentially heated walls in a strong magnetic field. Fluid Dynamics Research. 2004;**35**: 37-66

[28] Kahveci K, Oztuna S. MHD natural convection flow and heat transfer in a laterally heated partitioned enclosure. European Journal of Mechanics-B/ Fluids. 2009;**28**:744-752

[29] Kefayati GHR, Gorji M, Ganji DD, Sajjadi H. Investigation of Prandtl number effect on natural convection MHD in an open cavity by Lattice Boltzmann method. Engineering Computations. 2013;**30**:97-116

[30] Martinez D, Chen S, Matthaeus W. Lattice Boltzmann magneto hydrodynamics. Physics of Plasmas. 1994;(6):1850-1867

[31] Chaabane R, Askri F, Nasrallah SB. Parametric study of simultaneous transient conduction and radiation in a two-dimensional participating medium. Communications in Nonlinear Science and Numerical Simulation. 2011;**16**(10): 4006-4020

[32] Saha LK, Hossain MA, Gorla RSR. Effect of Hall current on the MHD laminar natural convection flow from a vertical permeable flat plate with uniform surface temperature. International Journal of Thermal Science. 2007;**46**:790-801

[33] Chaabane R, Askri F, Nasrallah SB. Analysis of two-dimensional transient conduction-radiation problems in an

anisotropically scattering participating enclosure using the Lattice Boltzmann method and the control volume finite element method. Journal of Computer Physics Communications. 2011;**182**(7): 1402-1413

[34] Lamsaadi M, Naimi M, Hasnaoui M, Mamou M. Natural convection in a vertical rectangular cavity filled with a non-Newtonian power law fluid and subjected to a horizontal temperature gradient. Numerical Heat Transfer Part A: Application. 2006;**49**:969-990

[35] Chaabane R, Askri F, Ben Nasrallah S. Application of the Lattice Boltzmann method to transient conduction and radiation heat transfer in cylindrical media. Journal of Quantitative Spectroscopy and Radiative Transfer. 2011;**112**(12):2013-2027

[36] Chaabane R, Askri F, Jemni A, Nasrallah SB. Numerical study of transient convection with volumetric radiation using an hybrid lattice Boltzmann BGK-control volume finite element method. Journal of Heat Transfer. 2017;**139**(9):092701-072017

[37] Chaabane R, Askri F, Jemni A, Nasrallah SB. Analysis of Rayleigh-Bénard convection with thermal volumetric radiation using Lattice Boltzmann formulation. Journal of Thermal Science and Technology. 2017; **12**(2)

[38] Mohamad AA. Applied Lattice Boltzmann Method for Transport Phenomena, Momentum, Heat and Mass Transfer. Calgary: Sure; 2007

[39] Succi S. The Lattice Boltzmann Equation for Fluid Dynamics and Beyond. Oxford, London: Clarendon Press; 2001

[40] Series RW, Hurle DTJ. The use of magnetic fields in semiconductor crystal growth. Journal of Crystal Growth. 1991;**133**:305-328

Chapter 5

Convection Flow of MHD Couple Stress Fluid in Vertical Microchannel with Entropy Generation

Abiodun A. Opanuga, Olasunmbo O. Agboola, Hilary I. Okagbue and Sheila A. Bishop

Abstract

Entropy generation of fully developed steady, viscous, incompressible couple stress fluid in a vertical micro-porous-channel in the presence of horizontal magnetic field is analysed in this work. The governing equations for the flow are derived, and nondimensionalised and the resulting nonlinear ordinary differential equations are solved via a rapidly convergent technique developed by Zhou. The solution of the velocity and temperature profiles are utilised to obtain the flow irreversibility and Bejan number. The effects of couple stresses, fluid wall interaction parameter (FSIP), effective temperature ratio (ETR), rarefaction and magnetic parameter on the velocity profile, temperature profile, entropy generation and Bejan number are presented and discussed graphically.

Keywords: microchannel, entropy generation, MHD, natural convection, differential transform method (DTM)

1. Introduction

In the last decades, the study of microchannel flows has become an important subject for researchers because of the reduction in the size of such devices which increases the dissipated heat per unit area. The effective performance of these devices is dependent on the temperature; as a result a comprehensive knowledge of such flow behaviours is required for accurate prediction of performance during the design process. These microfluidics have characteristic lengths of $1-100$ μm and are categorised by the dimensionless quantity called the Knudsen number (Kn). Researchers have shown that microchannel flows are influenced by several parameters of which velocity slip and temperature jump occurring at the solid-fluid interface in small-scale systems are the most important [2–4]. Velocity slip and temperature jump become more significant at higher Knudsen number. The latter boundary conditions are assumed at Knudsen number greater than 0.01 since below this value the classical Navier-Stokes equations is no longer valid.

The influence of velocity slip and temperature jump on microchannel flows has been extensively studied. Khadrawi and Al-Shyyab [5] obtained a close form solution of the effect of velocity slip and temperature jump on heat and fluid flow for axially moving micro-concentric cylinders. Chen and Tian [6] applied lattice Boltzmann numerical technique with Langmuir model for the velocity slip and temperature jump to investigate fluid flow and heat transfer between two horizontal parallel plates. The study suggested the application of Langmuir-slip model as an alternative for the Maxwell-slip model. Earlier on, Larrode et al. [7] in his work, slip-flow heat transfer in circular tubes, has stated the significance of fluid-wall interaction. Moreover, Adesanya [8] has analytically studied the effects of velocity slip and temperature jump on the unsteady free convective flow of heat-generating/heat-absorbing fluid with buoyancy force. The study concluded that an increase in the slip parameter enhanced fluid motion, while an increase in the temperature jump parameter increased fluid temperature. Aziz and Niedbalski [9] applied finite difference technique to investigate thermally developing microtube gas flow with respect to both radial and axial coordinates. The result indicated that increase in Knudsen number reduced local Nusselt number (Nu). Several investigations regarding microchannel flows have also been carried out by Jha and collaborators [10–12].

From the perspective of energy management, it has been established that thermal processes such as the microscale fluid flow and heat transfer modelling are irreversible. This implies that entropy generation which destroys the available energy leading to inefficiency of thermal designs exists. Therefore, the aim of this research endeavour is the minimisation of irreversibility associated with microchannel flow of couple stress fluid by applying the robust approach proposed by Bejan [13] and applied by numerous researchers such as Adesanya and collaborators [14–16], Adesanya and Makinde [17–18], Ajibade and Jha [19], Eegunjobi and Makinde [20–21], Das and Jana [22] and more recently Opanuga and his collaborators [23–26].

In this work, an efficient technique introduced by Zhou has been employed to construct the solutions of the velocity and temperature profiles. Due to the accuracy of this technique in handling numerous linear and nonlinear models of both ordinary and partial equations, it has gained wide applications by investigators over the last decades. Arikoglu and Ozkol [27] applied it to obtain the solution of difference equations. Biazar and Eslami [28] solved quadratic Riccati differential equation using this method. Agboola et al. [29–30] applied it to third-order ordinary differential equations and natural frequencies of a cantilever beam. Solutions of Volterra integral equation via this technique was obtained by Odibat [31], while Opanuga et al. [32] compared the method with Adomian decomposition method to find the solution of multipoint boundary-value problem and more recently in couple stress fluid model [33].

2. Problem formulation

A fully developed laminar, viscous, incompressible and electrically conducting couple stress fluid in a vertical parallel microchannel of width h is considered. The x-axis is such that it is vertically upward along the plates while the y-axis is taken normal to it. There is an asymmetric heating of the plates such that the hotter plate ($y = 0$) is maintained at temperature T_1, while the cooler plate ($y = h$) is at temperature T_2, ($T_1 > T_2$) and T_0 is the reference frame. Velocity slip and temperature jump are incorporated. Furthermore, induced magnetic field effect arising due to the motion of an electrically conducting fluid is taken into consideration (**Figure 1**). Following Cheng and Weng [34], the velocity slip and temperature jump are given as

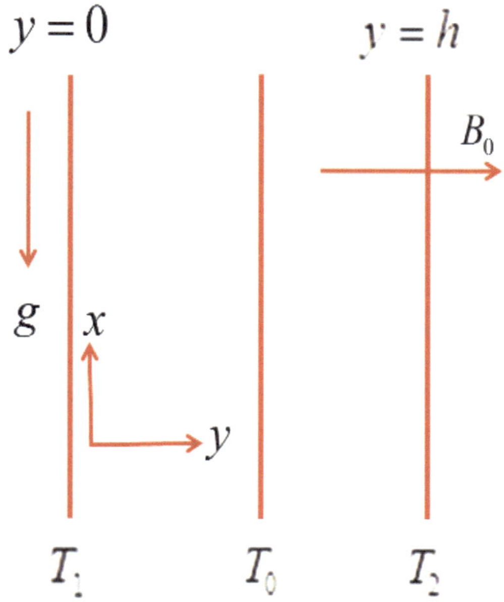

Figure 1.
Scheme of the vertical microchannel.

$$u_s = -\frac{2-\sigma_v}{\sigma_v}\lambda\frac{du}{dy'}\bigg|_{y'=h},\ T_s - T_w = -\frac{2-\sigma_t}{\sigma_t}\frac{2\gamma}{\gamma+1}\frac{\lambda}{\Pr}\frac{dT}{dy'}\bigg|_{y'=h} \tag{1}$$

While the governing equations for the flow are stated as [10]

$$\rho\nu_\circ\frac{du'}{dy'} = -\frac{dp}{dx'} + \mu\frac{d^2u'}{dy'^2} - \eta\frac{d^4u'}{dy'^4} + \rho g\beta^*(T - T_0) - \sigma B_0^2 u' \tag{2}$$

$$\rho c_p v_0\frac{dT'}{dy'} = k\frac{d^2T'}{dy'^2} + \mu\left(\frac{du'}{dy'}\right)^2 + \eta\left(\frac{d^2u'}{dy'^2}\right)^2 + \sigma B_0^2 u'^2 \tag{3}$$

The boundary conditions are

$$\begin{aligned} u' &= \frac{2-f_v}{f_v}\lambda\frac{du'}{dy'},\ \frac{d^2u'}{dy'^2} = 0,\ T' = T_2 + \frac{2-f_t}{f_t}\frac{2\gamma}{\gamma+1}\frac{\lambda}{\Pr}\frac{dT'}{dy},\ y = 0 \\ u' &= -\frac{2-f_v}{f_v}\lambda\frac{du'}{dy'},\ \frac{d^2u'}{dy'^2} = 0,\ T' = T_1 - \frac{2-f_t}{f_t}\frac{2\gamma}{\gamma+1}\frac{\lambda}{\Pr}\frac{dT'}{dy},\ y = h \end{aligned} \tag{4}$$

The dimensionless parameters for this flow are

$$\begin{aligned} &y = \frac{y'}{h},\ u = \frac{u'}{\nu_\circ},\ \theta = \frac{T' - T_0}{T_1 - T_0},\ \mathrm{Re} = \frac{\rho v_0 h}{\mu},\ G = \frac{-h^2}{\mu v_0}\frac{dp}{dx},\ a^2 = \mu\frac{h^2}{\eta}, \\ &\Pr = \frac{\nu\rho c_p}{k},\ Br = \frac{\mu v_0}{k(T_1 - T_0)},\ Bi_1 = \frac{\gamma_1 h}{k},\ Bi_2 = \frac{\gamma_2 h}{k},\ \nu = \frac{\mu}{\rho}, \\ &\Omega = \frac{T_1 - T_0}{T_0},\ H^2 = \frac{\sigma B_0^2 h^2}{\mu},\ \xi(T_2 - T_0) = T_1 - T_0,\ \beta_v = \frac{2-f_v}{f_v}, \\ &\beta_t = \frac{2-f_v}{f_v}\frac{2\gamma_s}{\gamma_s+1}\frac{1}{\Pr},\ \gamma_s = \frac{c_p}{c_v},\ Kn = \frac{\lambda}{h},\ \psi = \frac{\beta_t}{\beta_v},\ Ns = \frac{S_G T_0 h^2}{k(T_1 - T_0)} \end{aligned} \tag{5}$$

Using (5) in Eqs. (2)–(4) yields the boundary-value problems:

$$\mathrm{Re}\frac{du}{dy} = G + \frac{d^2u}{dy^2} - \frac{1}{a^2}\frac{d^4u}{dy^4} - H^2u + Gr\theta, \tag{6}$$

$$\frac{d^2\theta}{dy^2} = \mathrm{Re}\,p_r\frac{d\theta}{dy} - Br\left\{\left(\frac{du}{dy}\right)^2 + \frac{1}{a^2}\left(\frac{d^2u}{dy^2}\right)^2 + H^2u^2\right\}, \tag{7}$$

$$\begin{aligned} u &= \beta_v Kn\frac{du}{dy}, \frac{d^2u}{dy^2} = 0, \theta = \xi + \beta_v Kn\psi\frac{d\theta}{dy}, y = 0 \\ u &= -\beta_v Kn\frac{du}{dy}, \frac{d^2u}{dy^2} = 0, \theta = 1 - \beta_v Kn\psi\frac{d\theta}{dy}, y = 1 \end{aligned} \tag{8}$$

3. Method of solution

DTM is applied in this work to obtain the solution of the velocity and temperature profiles. The results are used to calculate the entropy generation and irreversibility ratio.

3.1 Differential transformation method (DTM)

Consider a function, $f(x)$. The differential transformation of the function $f(x)$ is defined as

$$F(k) = \frac{1}{k!}\left[\frac{d^kf(x)}{dx^k}\right]_{x=x_0} \tag{9}$$

where $f(x)$ is the given function and $F(k)$ is the transformed function which is also known as the spectrum of $f(x)$. The inverse differential transformation of $F(k)$ is given by

$$f(x) = \sum_{k=0}^{\infty} x^kF(k) \tag{10}$$

In practise, the function stated in Eq. (10) is usually represented by a finite series of the form:

$$f(x) = \sum_{k=0}^{n} x^kF(k) \tag{11}$$

where n is the size of the series (Arikoglu and Ozkol [27], Biazar and Eslami [28]).

To apply DTM to the problem in view, the basic properties of DTM, which are outlined in **Table 1**, are invoked in Eqs. (6)–(8). Doing this, one obtains the following recurrence relations:

$$F(k+4) = \frac{1}{a^2(k+4)!}\left[(k+1)(k+2)F(k+2) - \mathrm{Re}(k+1)F(k+1) - H^2(F(k)) + G\right] \tag{12}$$

Original function	Transformed function
$f(y) = u(y) \pm w(y)$	$F(k) = U(k) \pm W(k)$
$f(y) = \frac{d^n u(y)}{dy^n}$	$F(k) = \frac{(k+n)!}{k!} U(k+n)$
$f(y) = u^2$	$F(k) = \sum_{r=0}^{k} U(r)U(k-r)$
$f(y) = \left(\frac{du(y)}{dy}\right)^2$	$F(k) = \sum_{r=0}^{k} (r+1)(k-r+1)U(r+1)U(k-r+1)$
$f(y) = \left(\frac{d^2u(y)}{dy^2}\right)^2$	$F(k) = \sum_{r=0}^{k} (r+1)(r+2)(k-r+1)(k-r+2)U(r+2)U(k-r+2)$

Table 1.
Operations and properties of differential transform method.

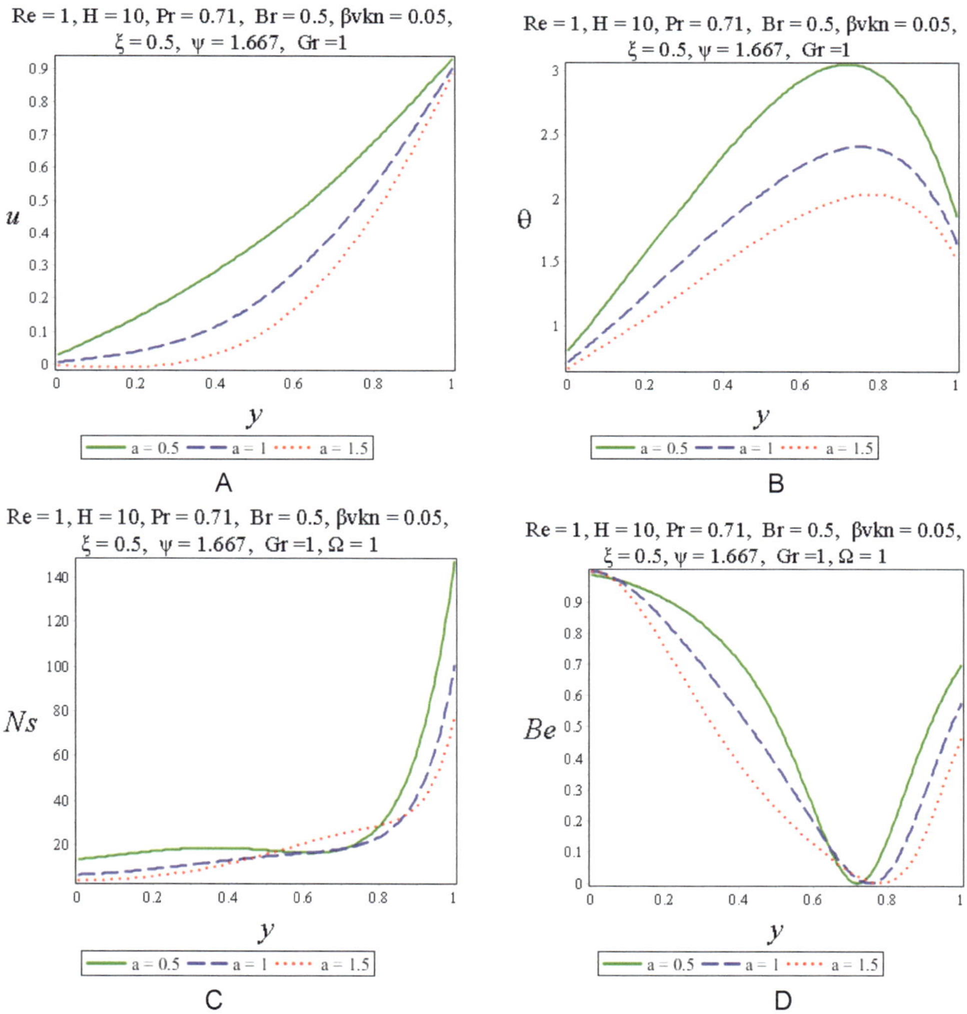

Figure 2.
A. Couple stress parameter vs. fluid velocity. B. Couple stress parameter vs. fluid temperature. C. Couple stress parameter vs. entropy generation. D. Couple stress parameter vs. Bejan number.

$$\Theta(k+2) = \frac{1}{(k+2)!}\left[\mathrm{RePr}(k+1)\Theta(k+1) - Br\left(\sum_{r=0}^{k}(r+1)F(r+1)(k-r+1)F(k-r+1) - \frac{Br}{a^2}\left(\sum_{r=0}^{k}(r+1)(r+2)F(r+2)(k-r+1)(k-r+2)F(k-r+2)\right)\right) - BrH^2\left(\sum_{r=0}^{k}F(r)F(k-r)\right)\right] \tag{13}$$

where $F(k)$ and $\Theta(k)$ are the transformed functions of $f(y)$ and $\theta(y)$, respectively. These are given by

$$f(y) = \sum_{k=0}^{\infty} y^k F(k), \theta(y) = \sum_{k=0}^{\infty} y^k \Theta(k). \tag{14}$$

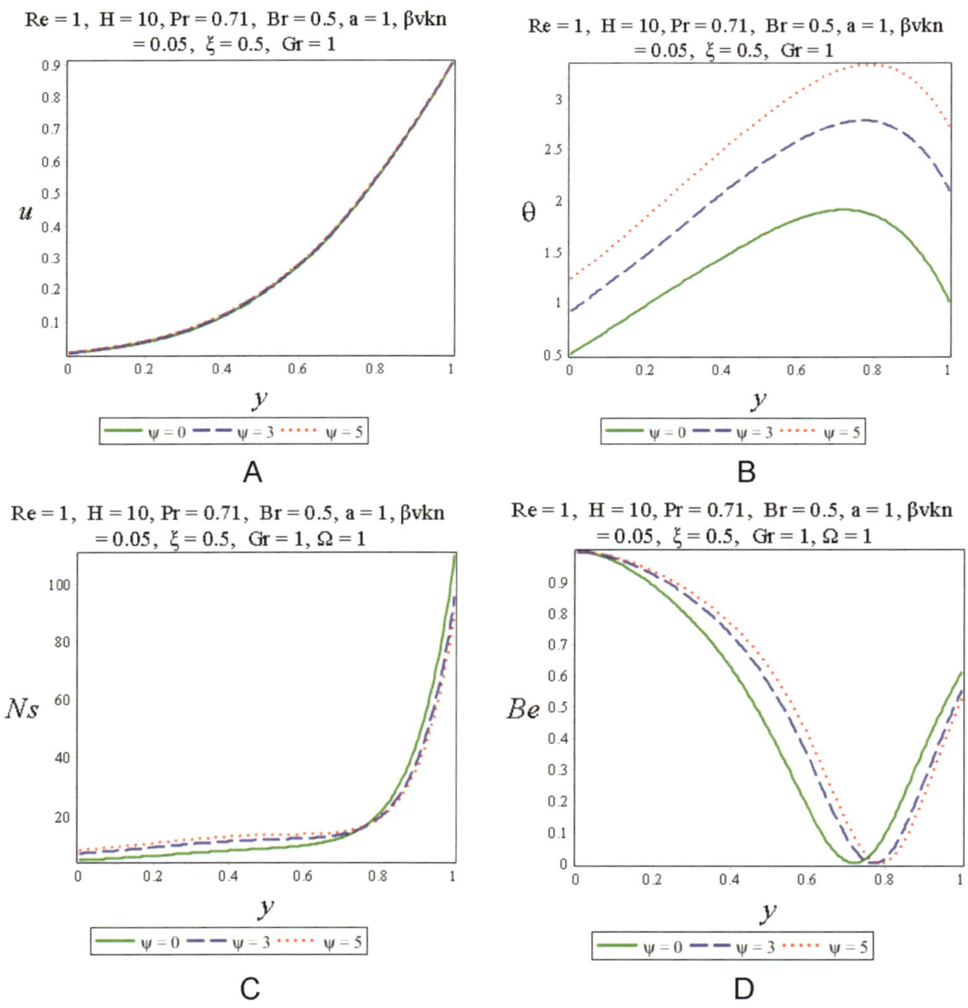

Figure 3.
A. Fluid wall interaction parameter vs. fluid velocity. B. Fluid wall interaction parameter vs. fluid temperature. C. Fluid wall interaction parameter vs. entropy generation. D. Fluid wall interaction vs. Bejan number.

Using the recursive relations of Eqs. (12) and (13), we can obtain the differential coefficients, $F(4), F(5), \ldots, F(n)$ and $\Theta(4), \Theta(5), \Theta(n)$ by setting $k = 0, 1, 2, \ldots$. The values of $F(j)$ for $j = 4, 5, \ldots$ and $\Theta(j)$ for $j = 2, 3, \ldots$ can now be evaluated in terms of $F(0), F(1), F(2), F(3), \Theta(0)$ and $\Theta(1)$. For convenience, the values of $F(0), F(1), F(2), F(3), \Theta(0)$ and $\Theta(1)$ are set as unknowns such as

$$\begin{aligned} &F(0) = a_1, \quad F(1) = a_2, \quad F(2) = a_3, \quad F(3) = a_4, \\ &\Theta(0) = b_1, \quad \Theta(1) = b_2. \end{aligned} \tag{15}$$

The values of $F(k)$ and $\Theta(k)$ for $k = 0, 1, \ldots$ are now substituted back into Eq. (14) to obtain the series solutions in the form:

$$f(y) = \sum_{k=0}^{n} y^k F(k), \quad \theta(y) = \sum_{k=0}^{n} y^k \Theta(k). \tag{16}$$

We next invoke the transformed form of boundary condition (8) on (16) to determine the values of all the unknown coefficients stated in (15). Coding

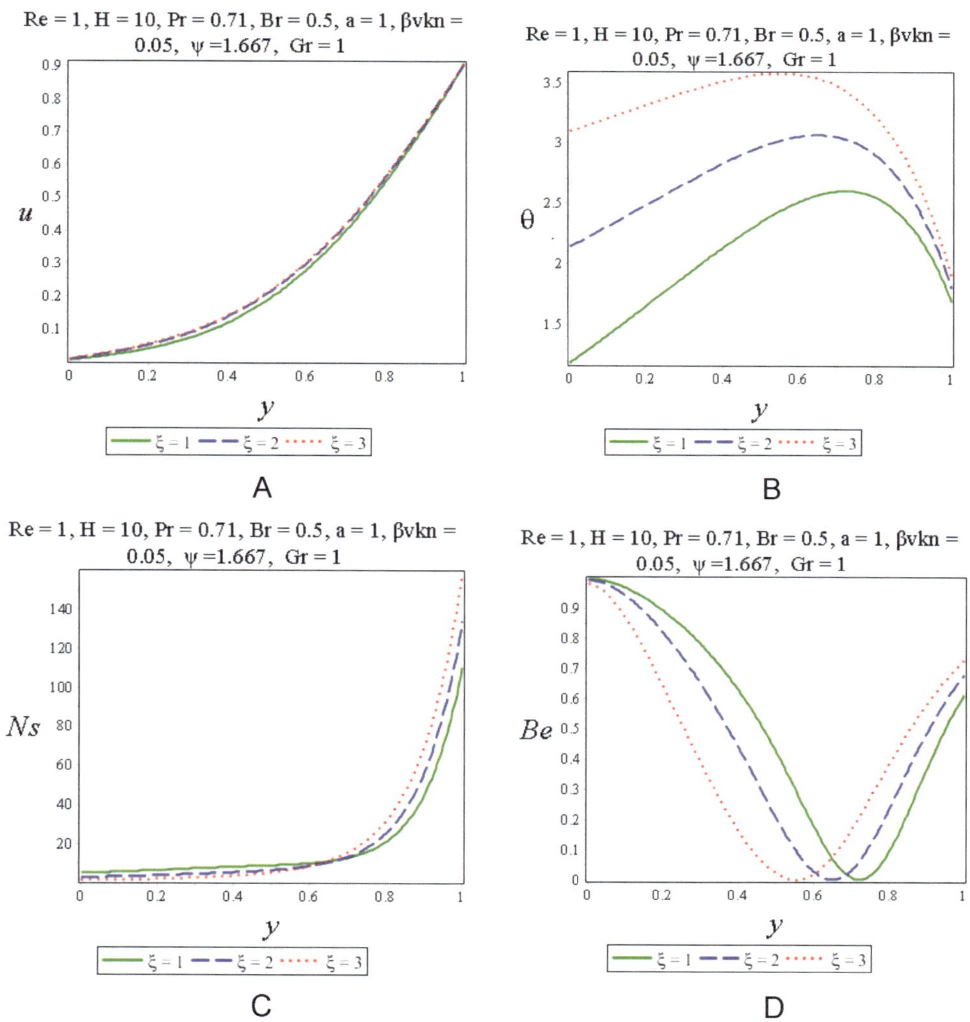

Figure 4.
A. Effective temperature ratio vs. fluid velocity. B. Effective temperature ratio vs. fluid temperature. C. Effective temperature ratio vs. entropy generation. D. Effective temperature ratio vs. Bejan number.

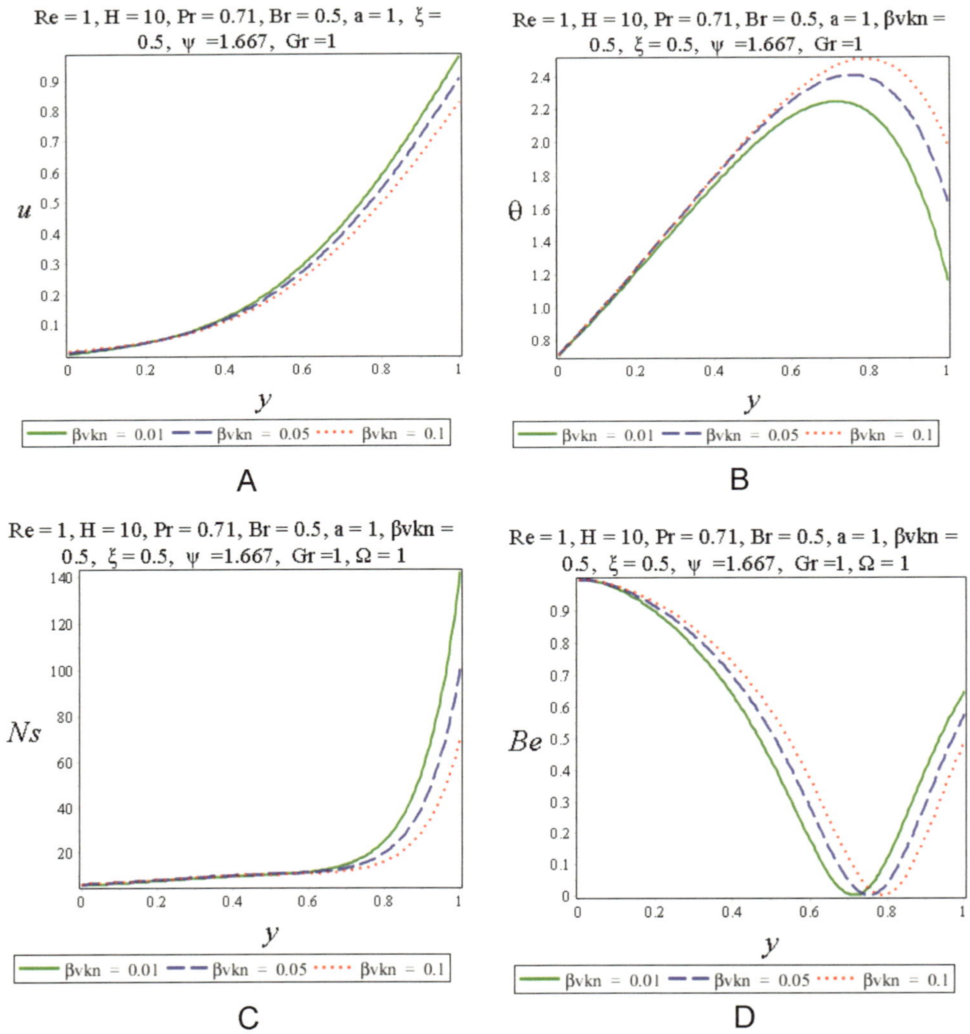

Figure 5.
A. Rarefaction vs. fluid velocity. B. Rarefaction vs. fluid temperature. C. Rarefaction vs. entropy generation. D. Rarefaction vs. Bejan number.

Eqs. (12)–(15) in symbolic Maple software yields the approximate solution. The results are presented in **Figures 2–6**.

To verify the accuracy of the results, the exact solution of the velocity profile (6) subject to the boundary conditions (8) at $\beta vKn = 0.05$, $Gr = 0$, $a = 1$, $\mathrm{Re} = 0.1$, $H = 1$ is obtained as.

$$u(y) = (-3.063639^{\backprime}e^{(0.866146^{\backprime}y)} + 2.057901^{\backprime}\,Cos\,[0.470472^{\backprime}y] + 1.^{\backprime}e^{(1.732291^{\backprime}y)}\,Cos\,[0.528272^{\backprime}y] + 1.808594^{\backprime}\,Sin\,[0.4704729^{\backprime}y] - 0.093418^{\backprime}e^{(1.732291^{\backprime}y)}\,Sin\,[0.528272^{\backprime}y]) \quad (17)$$

The above solution is compared with DTM solution as displayed in **Table 2**.

3.2 Analysis of entropy generation

The local entropy generation for the flow is given as Bejan [13]:

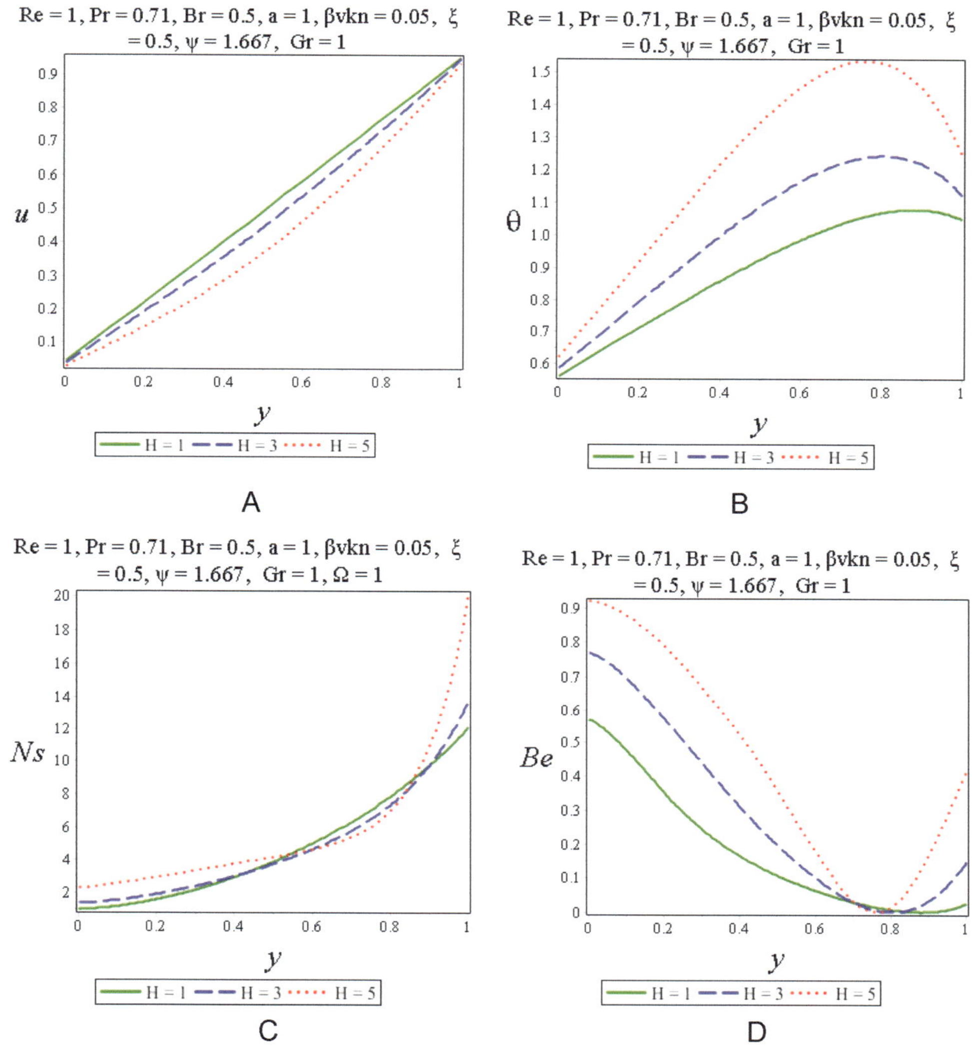

Figure 6.
A. Hartmann Number vs. fluid velocity. B. Hartmann Number vs. fluid temperature. C. Hartmann Number vs. entropy generation. D. Hartmann Number vs. Bejan number.

$$S_G = \frac{k}{T_0^2}\left(\frac{dT^*}{dy^*}\right)^2 + \frac{\mu}{T_0}\left(\left(\frac{du^*}{dy^*}\right)^2\right) + \frac{1}{a^2}\left(\left(\frac{d^2u}{dy^2}\right)^2\right) + \frac{\sigma B_0^2}{T_0}u^{*^2}. \tag{18}$$

The first term in Eq. (18) $\frac{k}{T_0^2}\left(\frac{dT^*}{dy^*}\right)^2$ is entropy generation due to heat transfer; the next term $\frac{\mu}{T_0}\left(\left(\frac{du^*}{dy^*}\right)^2\right)$ is entropy generation due to viscous dissipation, $\frac{1}{a^2}\left(\left(\frac{d^2u}{dy^2}\right)^2\right)$ and $\frac{\sigma B_0^2}{T_0}u^{*^2}$ are couple stress and magnetic entropy generation.

Using (4) in Eq. (18), the dimensionless form of entropy generation is written as

$$Ns = \left(\frac{d\theta}{dy}\right)^2 + \frac{Br}{\Omega}\left(\left(\frac{du}{dy}\right)^2\right) + \frac{Br}{\Omega a^2}\left(\left(\frac{d^2u}{dy^2}\right)^2\right) + \frac{BrM^2}{\Omega}u^2, \tag{19}$$

$\beta vKn = 0.05, Gr = 0, a = 1, Re = 0.1, H = 1$		
y	Exact	DTM
0	0.001872811813791	0.001872811814000
0.1	0.005546352540252	0.005546352541866
0.2	0.008819633339440	0.008819633342380
0.3	0.011379286267087	0.011379286271194
0.4	0.013003721837660	0.013003721842694
0.5	0.013560741309568	0.013560741315200
0.6	0.013006153436252	0.013006153442060
0.7	0.011383386215798	0.011383386221257
0.8	0.008824085393964	0.008824085398432
0.9	0.005549692555064	0.005549692557769
1	0.001873996484802	0.001873996484819

Table 2.
Comparison of the exact solution with the values of velocity (u).

where (S_G, Ns) are the dimensional and dimensionless entropy generation rates.
The ratio of heat transfer entropy generation (N_1) to fluid friction entropy generation (N_2) is represented as

$$\Phi = \frac{N_2}{N_1} \tag{20}$$

Alternatively, Bejan number gives the entropy generation distribution ratio parameter; it represents the ratio of heat transfer entropy generation (N_1) to the total entropy generation (N_s) due to heat transfer and fluid friction; it is defined as

$$Be = \frac{N_1}{N_s} = \frac{1}{1+\Phi}, \tag{21}$$

$$Be = \begin{cases} 0, N_2 \gg N_1 \\ 0.5, N_1 = N_2 \\ 1, N_2 \ll N_1 \end{cases} \tag{22}$$

Note that N_1 represents heat transfer irreversibility, while N_2 denotes irreversibility due to viscous dissipation, couple stresses and magnetic field.

4. Results and discussion

In this work, investigation has been conducted on fully developed, steady, viscous and incompressible flow of couple stress fluid in a vertical micro-porous-channel in the presence of magnetic field. Effects of couple stress parameter (a), fluid wall interaction parameter (ψ), effective temperature ratio (ETR) (ξ), rarefaction $(\beta_v Kn)$ and Hartmann number (H) are presented in this section. Reasonable intervals for the above parameters as used by Chen and Weng [32] are adopted in this investigation: $0 \le H \le 10$, $0 \le vKn \le 0.1$, $0 \le \psi \le 10$ and the selected reference values of $\beta vKn = 0.05$, $\ln = 1.667$. Furthermore, $0 \le \xi \le 5$ and $0 \le a \le 2$ with reference values of $a = 1$, $\xi = 0.5$.

4.1 Influence of couple stress parameter

Figure 2A illustrates the influence of couple stress parameter on fluid velocity. The plot shows a significant reduction in fluid velocity as the values of couple stress parameter increase. This observation is due to the increase in the dynamic viscosity of the fluid. In **Figure 2B**, it is illustrated that fluid temperature drops as couple stress parameter increases. It is depicted in **Figure 2C** that fluid entropy generation decreases at the microchannel walls as couple stress parameter increases. However, the effect is opposite near the middle of the microchannel. The same scenario is observed in **Figure 2D**; it is shown that Bejan number reduces in value at microchannel right wall which is an indication that fluid friction irreversibility is the major contributor to entropy generation as couple stress parameter increases.

4.2 Influence of fluid-structure interaction parameter

Figure 3A presents the effect of fluid-structure interaction parameter on fluid velocity. It is noted that FSIP does not have any significant effect on the slip velocity at the walls as well as the microchannel. However, **Figure 3B** reveals that fluid temperature is enhanced as FSIP increases. Response to the enhancement in fluid temperature in **Figure 3B** is the rise in fluid entropy generation at the hotter wall of the microchannel, while entropy generation is reduced at the cooler wall as depicted in **Figure 3C**. In **Figure 3D**, it is noticed that Bejan number increases at the hotter wall of the microchannel, while it reduces at the cooler wall. The implication of the latter is that heat transfer irreversibility is the cause of entropy generation at the hotter wall, while on the other hand, fluid friction irreversibility is the major contributor at the cooler wall of the microchannel.

4.3 Influence of effective temperature ratio

Next is the response of fluid velocity, fluid temperature and entropy generation to variation in effective temperature ratio. In **Figure 4,** it is observed that velocity is accelerated slightly except towards the microchannel centre and plate $y = 0$. In **Figure 4B,** fluid temperature is significantly enhanced at increasing values of effective temperature ratio. The effect of this is noticed in **Figure 4C** with an increase in entropy production approaching the microchannel plate $y = 1$. Furthermore, Bejan number rises approaching the plate $y = 1$ but reduces towards $y = 0$ in **Figure 4D**. It is then concluded that fluid friction irreversibility is dominant at plate $y = 0$, while heat transfer irreversibility is dominant at plate $y = 1$.

4.4 Influence of rarefaction

In **Figure 5A,** the effect of rarefaction parameter on fluid velocity is presented. The plot indicates that rarefaction parameter increases and fluid velocity at plate $y = 0$ is not significant; however, it is decelerated towards the microchannel plate $y = 1$. Fluid temperature is considerably enhanced at plate $y = 1$ for different values of rarefaction as displayed in **Figure 5B**. **Figures 5C** and **5D** presents similar results at plate $y = 1$. Both entropy generation and Bejan number reduce as the values of rarefaction parameter increase. The implication of this observation is that entropy generation at microchannel wall $y = 1$ is a consequence of viscous dissipation.

4.5 Influence of Hartmann number

Finally, the response of fluid velocity, temperature, entropy generation and Bejan number to variation in Hartmann number is presented in **Figure 6**. In **Figure 6A,** fluid velocity is found to have decelerated within the microchannel region. The observed reduction in the motion of fluid is attributed to the presence of applied magnetic field which usually induces a resistive type of force known as Lorentz force. Also the presence of Ohmic heating in the flow significantly enhanced fluid temperature as depicted in **Figure 6B**. **Figures 6C** and **6D** displays increase in fluid entropy generation and Bejan number at the microchannel walls as Hartmann number increases from 1 to 5. It is deduced that fluid entropy generation is induced by heat transfer irreversibility.

5. Conclusions

Entropy generation of fully developed steady, viscous, incompressible couple stress fluid in a vertical micro-porous-channel in the presence of magnetic field is analysed in this work. The equations governing the fluid flow are solved via an efficient technique proposed by Zhou. Then fluid entropy generation and Bejan number are calculated by the results obtained. This work reduces to Chen and Weng [34] when Hartmann number, couple stress parameter, entropy generation and Bejan number are neglected $(H \to 0, a \to 0, Ns \to 0, Be \to 0)$. Furthermore, it agrees with Jha and Aina [10] in the absence of couple stress parameter, entropy generation and Bejan number $(a \to 0, Ns \to 0, Be \to 0)$. The present study is significant in the cooling of microchannel devices and conservation of useful energy. The following conclusions are made based on the results above:

1. Couple stress parameter reduces fluid velocity, velocity slip, temperature and entropy generation.

2. Increase in fluid-structure interaction parameter increases fluid temperature. However, entropy generation enhances at the hotter wall and reduces at the cooler region of the microchannel. Furthermore, fluid irreversibility is enhanced at the hot wall and increases at the cold wall.

3. Effective temperature ratio slightly enhances fluid velocity and velocity slip, raising fluid temperature significantly.

4. An increase in the values of rarefaction mainly reduces fluid velocity and velocity slip, increases fluid temperature and reduces entropy generation.

5. Hartmann number decreases fluid velocity and velocity slip, while the temperature is enhanced considerably. Entropy generation and Bejan number are enhanced at microchannel walls.

Acknowledgements

Authors appreciate the funding provided by Covenant University, Ota.

Nomenclature

B_0	uniform magnetic field
u	fluid velocity
h	channel width
f_t, f_v	thermal and tangential momentum accommodation coefficients, respectively
Cp	specific heat capacity
Re	Reynolds number
a	couple stress parameter
k	thermal conductivity
Kn	Knudsen number
H	Hartmann number
Pr	Prandtl number
T	temperature of fluid
T_0	reference temperature
Br	Brinkman number
E_G	local volumetric entropy generation rate
Be	Bejan number
Cv	specific heats at constant volume
Ns	dimensionless entropy generation parameter

Greek letters

ρ	fluid density
β_t, β_v	dimensionless variables
γ_s	ratio of specific heat
μ	dynamic viscosity
ξ	effective temperature ratio
σ	electrical conductivity
Ω	temperature difference
η	fluid particle size effect due to couple stresses
ψ	fluid wall interaction parameter

Author details

Abiodun A. Opanuga*, Olasunmbo O. Agboola, Hilary I. Okagbue and Sheila A. Bishop
Department of Mathematics, Covenant University, Ota, Nigeria

*Address all correspondence to: abiodun.opanuga@covenantuiversity.edu.ng

IntechOpen

© 2018 The Author(s). Licensee IntechOpen. This chapter is distributed under the terms of the Creative Commons Attribution License (http://creativecommons.org/licenses/by/3.0), which permits unrestricted use, distribution, and reproduction in any medium, provided the original work is properly cited. CC BY

References

[1] Agboola OO, Opanuga AA, Okagbue HI, Bishop SA, Ogunniyi PO. Analysis of hall effects on the entropy generation of natural convection flow through a vertical microchannel. International Journal of Mechanical Engineering and Technology. 2008;**9**(8):712-721

[2] Arkilic EB, Schmidt MA, Breuer KS. Gaseous slip flow in long microchannels. Journal of Microelectromechanical Systems. 1997;**6**:167-178

[3] Sofonea V, Sekerka RF. Diffuse-reflection boundary conditions for a thermal lattice Boltzmann model in two dimensions: Evidence of temperature jump and slip velocity in microchannels. Physical Review E – Statistical Nonlinear and Soft Matter Physics. 2005;**71**:1-10 http://dx.doi.org/10.1103/PhysRevE.71.066709

[4] Lv Q, Liu X, Wang E, Wang S. Analytical solution to predicting gaseous mass flow rates of microchannels in a wide range of Knudsen numbers. Physical Review E. 2013;**88**:013007 http://dx.doi.org/10.1103/PhysRevE.88.013007

[5] Khadrawi AF, Al-Shyyab A. Slip flow and heat transfer in axially moving micro-concentric cylinders. International Communications in Heat and Mass Transfer. 2010;**37**:1149-1152 http://dx.doi.org/10.1016/j.icheatmasstransfer.2010.06.006

[6] Chen S, Tian Z. Simulation of thermal micro-flow using lattice Boltzmann method with Langmuir slip model. International Journal of Heat and Fluid Flow. 2010;**31**:227-235 http://dx.doi.org/10.1016/j.ijheatfluidflow.2009.12.006

[7] Larrode FE, Housiadas C, Drossinos Y. Slip-flow heat transfer in circular tubes. International Journal of Heat and Mass Transfer. 2000;**43**:2669-26680

[8] Adesanya SO. Free convective flow of heat generating fluid through a porous vertical channel with velocity slip and temperature jump. Ain Shams Engineering Journal. 2015;**6**:1045-1052. DOI: http://dx.doi.org/10.1016/j.asej.2014.12.008

[9] Aziz A, Niedbalski N. Thermally developing microtube gas flow with axial conduction and viscous dissipation. International Journal of Thermal Sciences. 2011;**50**:332-340 http://dx.doi.org/10.1016/j.ijthermalsci.2010.08.003

[10] Jha BK. Aina B. MHD natural convection flow in a vertical porous microchannel formed by nonconducting and conducting plates in the presence of induced magnetic field. Heat Transfer Research. 2017;**48**(15):1-24

[11] Jha BK, Aina B, Joseph SB. Natural convection flow in a vertical microchannel with suction/injection. Proceedings of the Institution of Mechanical Engineers, Part E: Journal of Process Mechanical Engineering. 2014;**228**(3):171-180. DOI: 10.1177/0954408913492719

[12] Jha BK, Aina B, Ajiya AT. MHD natural convection flow in a vertical parallel plate microchannel. Ain Shams Engineering Journal. 2015;**6**:289-295

[13] Bejan A. Entropy Generation through Heat and Fluid Flow. New York: Wiley; 1982

[14] Adesanya SO, Falade JA, Jangili S, Beg OA. Irreversibility analysis for reactive third-grade fluid flow and heat transfer with convective wall cooling. Alexandria Engineering Journal. 2017;**56**:153-160

[15] Adesanya SO, Kareem SO, Falade JA, Arekete SA. Entropy generation analysis for a reactive couple stress fluid

flow through a channel saturated with porous material. Energy. 2015;**93**: 1239-1245

[16] Jangili S, Adesanya SO, Falade JA, Gajjela N. Entropy generation analysis for a radiative micropolar fluid flow through a vertical channel saturated with non-Darcian porous medium. International Journal of Applied and Computational Mathematics. **3**(4): 3759-3782. DOI: 10.1007/s40819-017-0322-8

[17] Adesanya SO, Makinde OD. Irreversibility analysis in a couple stress film flow along an inclined heated plate with adiabatic free surface. Physica A: Statistical Mechanics and its Applications. 2015;**432**:222-229

[18] Adesanya SO, Makinde OD. Effects of couple stresses on entropy generation rate in a porous channel with convective heating. Computational & Applied Mathematics. 2015;**34**:293-307. DOI: 10.1007/s40314-014-0117-z

[19] Ajibade AO, Jha BK, Omame A. Entropy generation under the effect of suction/injection. Applied Mathematical Modelling. 2011;**35**:4630-4646

[20] Eegunjobi AS, Makinde OD. Effects of Navier slip on entropy generation in a porous channel with suction/injection. Journal of Thermal Science and Technology. 2002;**7**(4): 522-535

[21] Eegunjobi AS, Makinde OD. Combined effect of buoyancy force and Navier slip on entropy generation in a vertical porous channel. Entropy. 2012; **14**:1028-1044. DOI: 10.3390/e14061028

[22] Das S, Jana RN. Entropy generation due to MHD flow in a porous channel with Navier slip. Ain Shams Engineering Journal. 2014;**5**:575-584

[23] Opanuga AA, Okagbue HI, Agboola OO, Imaga OF. Entropy generation analysis of buoyancy effect on hydromagnetic poiseuille flow with internal heat generation. Defect and Diffusion Forum. 2017;**378**:102-112

[24] Opanuga AA, Gbadeyan JA, Iyase SA. Second law analysis of hydromagnetic couple stress fluid embedded in a non-Darcian porous medium. IAENG International Journal of Applied Mathematics. 2017;**47**(3): 287-294

[25] Opanuga AA, Okagbue HI, Agboola OO. Irreversibility analysis of a radiative MHD Poiseuille flow through porous medium with slip condition. Proceedings of The World Congress on Engineering 2017; July 5–7; London, U.K

[26] Opanuga AA, Bishop SA, Okagbue HI, Agboola OO. Hall current and joule heating effects on flow of couple stress fluid with entropy generation. Engineering, Technology & Applied Science Research. 2008;**8**(3):2923-2930

[27] Arikoglu A. Ozkol solution of difference equations by using differential transform method. Applied Mathematics and Computation. 2006; **173**(1):126-136. http://dx.doi.org/10.1016/j.amc.2005.06.013

[28] Biazar J, Eslami M. Differential transform method for quadratic Riccati differential equation. International Journal of Nonlinear Science. 2010;**9**(4): 444-447

[29] Agboola OO, Gbadeyan JA, Opanuga AA, Agarana MC, Bishop SA, Oghonyon JG. Variational iteration method for natural frequencies of a cantilever beam with special attention to the higher modes. Proceedings of The World Congress on Engineering 2017; July 5–7 London; UK

[30] Agboola OO, Opanuga AA, Gbadeyan JA. Solution of third order

ordinary differential equations using differential transform method. Global Journal of Pure and Applied Mathematics. 2015;**11**(4):2511-2517

[31] Odibat Z. Differential transform method for solving Volterra integral equation with separable kernels. Mathematical and Computer Modelling. 2008;**48**:144-149. http://dx.doi.org/10.1016/j.mcm.2007.12.022

[32] Opanuga AA, Agboola OO, Okagbue HI. Approximate solution of multipoint boundary value problems. Journal of Engineering and Applied Sciences. 2015;**10**(4):85-89

[33] Opanuga AA, Okagbue HI, Agboola OO, Bishop SA. Second law analysis of ion slip effect on MHD couple stress fluid. International Journal of Mechanics. 2018;**12**:96-101

[34] Chen CK, Weng HC. Natural convection in a vertical microchannel. Journal of Heat Transfer. 2005;**127**:1053-1056

Chapter 6

Free Convection in a MHD Open Cavity with a Linearly Heated Wall Using LBM

Raoudha Chaabane

Abstract

This chapter examines the magneto-hydrodynamic (MHD) free convection in a square enclosure filled with liquid gallium subjected to a [transverse magnetic field] in-plane magnetic field. First, the side vertical walls of the cavity have spatially varying linearly temperature distributions. The bottom wall is uniformly heated and the upper wall is adiabatic. The second configuration is an open enclosure heated linearly from the left wall. Lattice Boltzmann method (LBM) is applied in order to solve the coupled equations of flow and temperature fields. This study has been carried out for Ra of 10^5. The results show that the heat transfer rate decreases with an increase of the Ha number which is a widely solicited result in different engineering applications. Streamlines, isotherm counters and Nu numbers are displayed and discussed. A good stability is observed for all studied cases employing an in-house code. The obtained results show that the free-convection heat transfer in the open MHD enclosure is enhanced and it is greater to that of a uniformly heated wall.

Keywords: MHD, open cavity, free convection, LBM, heat transfer

1. Introduction

Free convection in open cavities is an important phenomenon in engineering systems because of this it have received a considerable attention due to their applications in various industries of high-performance insulation for buildings, injection molding chemical catalytic reactors, packed sphere beds, grain storage, float glass production, air-conditioning in rooms, cooling of electronic devices, and geophysical problems. Extensive research studies using various numerical simulations were conducted into free convection of open enclosures [1–5]. The latter works have acquired a basic understanding of free convection flows and heat transfer characteristics in an open enclosure with non-conducting fluid. However, in most studies, one vertical wall of the enclosure is cooled and another one heated while the remaining top and bottom walls are well insulated. Recently, MHD free convection flows have attracted attention since can be are encountered in numerous problems with industrial and technological interest, covering a wide range of basic sciences such as nuclear engineering, fire research, crystal growth, astrophysics and metallurgy [6]. Besides, the process of manufacturing materials in industrial problems and microelectronic heat transfer devices involves an electrically conducting fluid subjected to magnetic field. In this case the fluid experiences a Lorentz force reducing the flow velocities which affects the heat transfer rate [7, 8]. In MHD free convection flows, the balance is achieved by inertial, viscous, electromagnetic and buoyancy forces, rendering the solution more complicated. Other useful research works had been conducted to simulate the MHD free convection under different conditions [9–18]. Kahveci and Oztuna [19] investigated MHD free convection flow and heat transfer in a laterally heated partitioned enclosure. They showed that the horizontal magnetic field is more effective in damping convection than the vertical one. Furthermore, Kefayati et al. [20] investigated the effect of Ra number on free convection MHD in an open cavity. Nonetheless several investigations in MHD free convection inside open enclosure with linearly heated wall have been carried out. In addition to conventional free convection in enclosures with uniform thermic boundaries, recent attention has been intensively focused on the cases with mixed boundary conditions on the walls of an open cavity [21–24].

On the other hand, the Lattice Boltzmann Method (LBM) is a new method for simulating fluid flow and modeling physics in fluids. It has been applied extensively within the last decade. Based on kinetic theory for simulating fluid flows and modeling the physics in fluids [25–32], it becomes a powerful, effective and easy numerical method, for simulation of complex flow problems with different boundary conditions [33, 34].

In comparison with the conventional CFD methods, the LBM is based on the microscopic models, mesoscopic kinetic equations. The macroscopic dynamics of a fluid is the result of collective behavior of many microscopic particles in the system. The LBM has been proved to recover the Navier-Stokes equation by using the Chapman-Enskog expansion. The major advantage of this method is its explicit feature of the governing equation, easy for parallel computation and easy for implementation of irregular boundary conditions. To our best knowledge, no previous study on effects of linearly heated wall on free convection in an open MHD cavity with the LBM had already been studied so far. The main aim of this chapter is to study effects of linearly heated wall on flow field and temperature distribution in an open MHD cavity filled with liquid gallium and also to highlight the ability of the LBM for solving problems with various complex boundary conditions. The effects of Ra number on streamlines, isotherms and the Nusselt number are investigated.

2. Mathematical formulation

2.1 Problem statement

The considered model is shown in **Figure 1**. It displays a two-dimensional open cavity with side length of H. At first case the left vertical is maintained at high temperature (T_H). Whereas at the second case, the vertical left wall is linearly heated. An external cold air enters into the enclosure from the east opening boundary, while the fluid is correlated with open boundary at constant temperature (T_C). The horizontal walls are insulated and impermeable to mass transfer. The bottom wall is at high temperature TH and the top one is adiabatic. The open cavity is filled with liquid gallium with Pr number of 0.025. The gravitational acceleration acts downward. The uniform external magnetic field with a constant magnitude B_0 is applied in the x-direction (transverse field). It is assumed that the induced magnetic field produced by the motion of an electrically conducting fluid is negligible compared to the applied magnetic field. Thermo-physical properties of the fluid are assumed to be constant, and the density variation in the buoyancy force term is handled by the Boussinesq approximation. The flow is two-dimensional, laminar

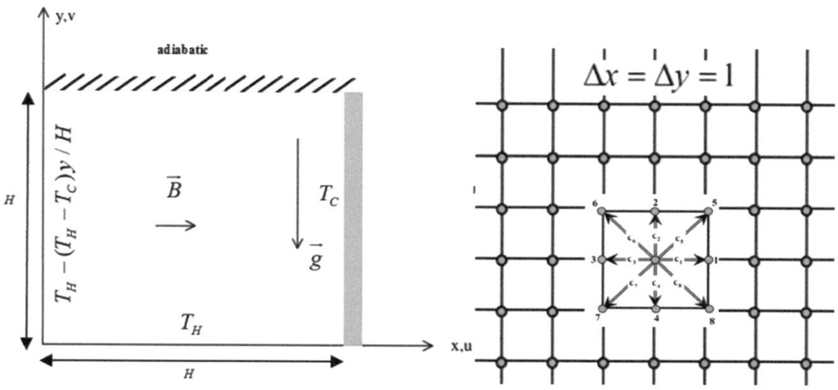

Figure 1.
Geometry of the present study with imposed boundary conditions and standard D_2Q_9 velocity mode.

and incompressible; in addition, it is assumed that the viscous dissipation and Joule heating are neglected.

2.2 Brief introduction to LBM

A scheme of the standard D2Q9 (**Figure 1**) for flow and for temperature, LBM method are used in this work [33–34] hence only brief discussion will be given in the following paragraphs, for completeness. Therefore, classic governing equations for MHD free convection are written in terms of the macroscopic variable depending on position x, y as:

Continuity equation

$$\frac{\partial u}{\partial x} + \frac{\partial v}{\partial y} = 0 \tag{1}$$

Momentum equations

$$\rho\left(u\frac{\partial u}{\partial x} + v\frac{\partial u}{\partial y}\right) = -\frac{\partial p}{\partial x} + \mu\left(\frac{\partial^2 u}{\partial x^2} + \frac{\partial^2 u}{\partial y^2}\right) + F_x \tag{2}$$

$$\rho\left(u\frac{\partial v}{\partial x} + v\frac{\partial v}{\partial y}\right) = -\frac{\partial p}{\partial y} + \mu\left(\frac{\partial^2 v}{\partial x^2} + \frac{\partial^2 v}{\partial y^2}\right) + F_y \tag{3}$$

Energy equation

$$u\frac{\partial T}{\partial x} + v\frac{\partial T}{\partial y} = \alpha\left(\frac{\partial^2 T}{\partial x^2} + \frac{\partial^2 T}{\partial y^2}\right) \tag{4}$$

Where $\nu = \mu/\rho$ is the kinematic viscosity, F_x and F_y are the body forces at horizontal and vertical directions respectively and they are defined as follows [25]:

$$F_x = R\left(v \sin\gamma \cos\gamma - u \sin^2\gamma\right) \tag{5}$$

$$F_y = R\left(u \sin\gamma \cos\gamma - v \cos^2\gamma\right) + \rho g \beta (T - T_m) \tag{6}$$

Where the Ha number is defined as

$$Ha = L B_x \sqrt{\sigma/\mu} \tag{7}$$

In the LBM the total force is:

$$F = F_x + F_y \tag{8}$$

$$F_x = 3w_k\rho\left[R(v\sin\gamma\cos\gamma) - u\sin^2\gamma)\right] \tag{9}$$

$$F_y = 3w_k\rho\left[g_y\beta(T - T_m) + R(u\sin\gamma\cos\gamma) - v\cos^2\gamma\right] \tag{10}$$

Where $R = \mu Ha^2$ and γ is the direction of the magnetic field.

The functional density, velocity and temperature in the mesoscopic approach are:

$$\rho(\mathbf{r},t) = \sum_k f_k(\mathbf{r},t) \tag{11}$$

$$\mathbf{u}(\mathbf{r},t) = \sum_k \mathbf{c}_k f_k(\mathbf{r},t)/\rho(\mathbf{r},t) \tag{12}$$

$$T = \sum_k g_k(\mathbf{r},t) \tag{13}$$

2.3 Method of solution

To ensure that the <code simulates an approximated incompressible regime>, the characteristic velocity of the flow ($V = \sqrt{g\beta_T H(T_w - T_\infty)}$) at the free convection regime must be compared with the fluid speed of sound c_s. In the present study, the characteristic velocity is selected as 0.1 of sound speed. By fixing Ra number $Ra = \beta g_y H^3 \mathrm{Pr}(T_H - T_C)/\nu^2$, Pr number $\mathrm{Pr} = \alpha\nu$ and Mach number ($Ma = u/c$), the kinematic viscosity $\nu = Ma\,Ha\,c_s\sqrt{\mathrm{Pr}/Ra}$ and thermal diffusivity α are deduced. Nusselt number Nu is one of the most important dimensionless parameters in describing the convective and thermic heat transport. For the example, the local Nusselt number and the average value at the left side wall are calculated as;

$$Nu_y = -\frac{H}{T_H - T_C}\frac{\partial T}{\partial x}\bigg|_{x=0} \tag{14}$$

$$Nu_{av} = \frac{1}{H}\int_0^H Nu_y dy \tag{15}$$

2.4 Algorithm of the numerical procedure

In this section, we present the steps that must be executed for the numerical implementation:

1. Input constant parameters.

2. Calculate the relaxation times.

3. Initialize the temperature, and velocity fields for the entire computational domain and the equilibrium distribution functions.

4. Calculate the current distribution function for propagation of temperature and velocity.

5. Propagate all current distribution functions to the neighbor lattices in all directions.

6. Impose the boundary conditions.

7. Calculate the current temperature and velocity field for the entire computational domain and the current equilibrium distribution functions using the current temperature and velocity fields.

8. The error convergence is checked, if the error does not satisfies the convergence criterion, go to step 4.

3. Results and discussion

For all further numerical cases, both for flow and temperature fields, the computations are based on an iterative LBM used for obtaining the numerical simulations. In order to validate the numerical code, the pure free convection, in a square open cavity with insulated horizontal walls filled with air was solved, and the results were compared with the numerical ones reported by reference [20]. The present findings are obtained with an extended computational domain. In **Figure 2**, we depict the comparison of the streamlines and isotherms of the present study with the prediction of [20]. Obtained results demonstrate that a good agreement was achieved. Also, in **Figure 3**, a comparison of isotherms and streamlines demonstrate a good agreement with previous work [7] where the considered matter is MHD free convection in a closed cavity with linearly heated west wall which is filled with liquid gallium (Pr = 0.025) and Ha = 50. The present chapter extend the study to deal with free convection in MHD open cavity with constant or linearly heated west wall which is filled with liquid gallium with Pr = 0.025 and Ha = 50 for Ra = 10^5.

The heat transfer rate is highlighted within the local and average Nusselt number in the case of linearly heated side walls. **Figure 4b** illustrates the variation of the local Nusselt number at the bottom wall of the cavity Nu_b with the horizontal normalized distance x/X for the case of linearly heated side walls with Ha = 50, Pr = 0.025 and Ra = 10^5. The Nusselt local value at the bottom is equal to unity at the

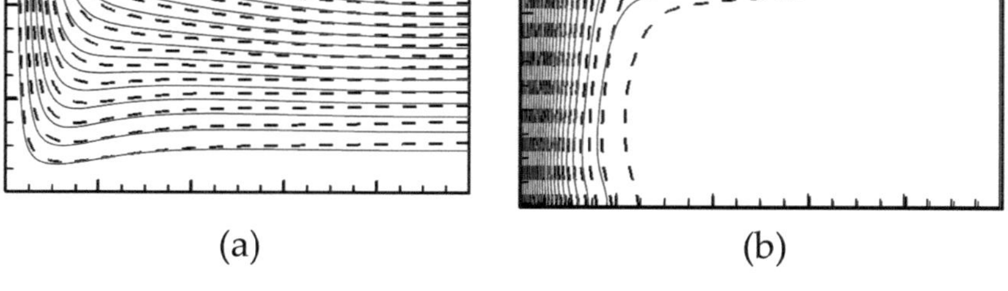

(a) (b)

Figure 2.
Comparison of the steady state streamlines (a) and isotherms (b) at Pr = 0.71 for Ha = 0 and Ra = 10^5 between [20] (continuous lines) and the present work (dashed lines).

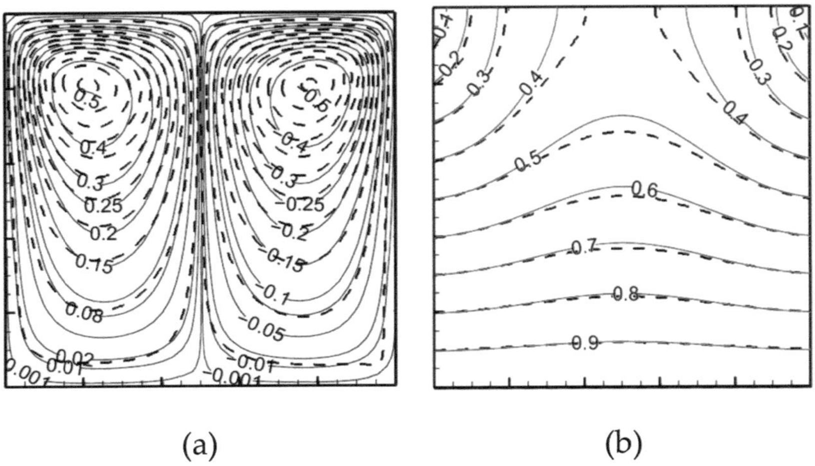

Figure 3.
Of the steady state fluid velocity streamlines (a) and isotherms (b) of linearly heated side walls MHD cavity for Ha = 50, Pr = 0.025 and Ra = 10^5 between [7] (continuous lines) and the present work (dashed lines). Frame ranges [−0.001, 0.5] (a), frame ranges [0, 1] (b).

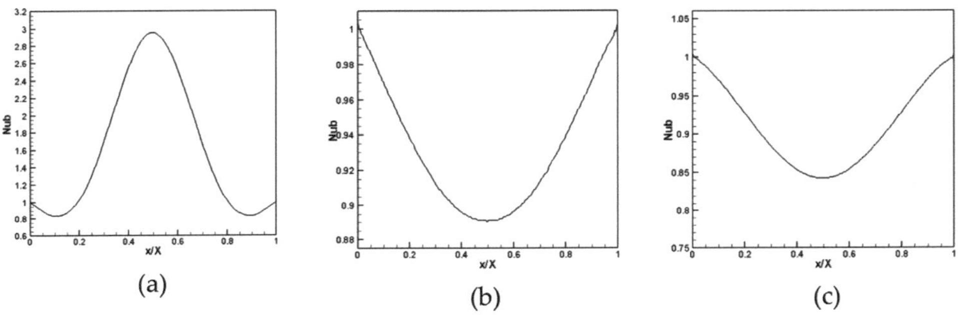

Figure 4.
Variation of bottom Nusselt number with distance at bottom wall in the case of linearly heated side walls (Pr = 0.025 and Ra = 10^5) for different Ha numbers. (a) Ha = 0, (b) Ha = 50 and (c) Ha = 150.

left (x/X = 0) and right (x/X = 1). This is an expected result due to the linearly heated side walls. The latter is associated with the symmetry in the thermic boundary conditions of the side walls of the cavity. **Figure 4a–c** shows clearly that the local Nusselt number at the bottom wall Nu_b exhibits an oscillatory behavior with the horizontal distance x/X and that it is exactly symmetric about the centerline of the bottom wall. The Variation of local Nusselt number with distance at bottom wall in the case of linearly heated side walls (Pr = 0.025 and Ra = 10^5) is depicted for different Ha numbers. **Figure 5b** depicts the variation of the local Nusselt number at the linearly heated side walls of the cavity Nu sides with the vertical distance y/Y for Ha = 50 while **Figure 5a** depicts the variation of the local Nusselt number at the linearly heated side walls of the cavity Nu sides with the vertical distance y/Y for Ha = 0. In general, the values of Nu sides increase as the vertical distance y/Y increases reaching maximum at y/Y = 1. Also, for Ha = 0, the distributions of Nu sides along the vertical distance y/Y oscillate close to lower end of the vertical side wall and continues with an increasing trend in its upper end. This behavior is believed to be related the feature represented by two pairs of symmetric cells with clockwise and counter-clockwise rotations [7].

Figures 6 and 7 display steady-state contour maps for the isotherms and the streamlines contours at various Ha numbers (0, 50 and 100) for Pr = 0.025 and

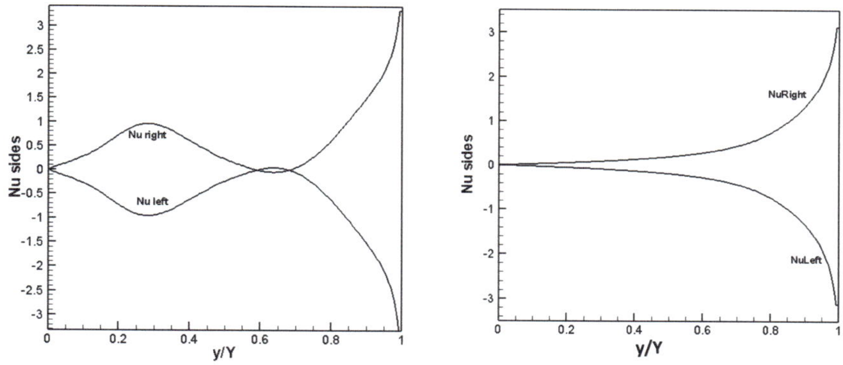

(a) (b)

Figure 5.
Variation of local Nusselt number with distance at side wall for linearly heated side walls (Pr = 0.025 and Ra = 10^5). (a) Ha = 0 and (b) Ha = 50.

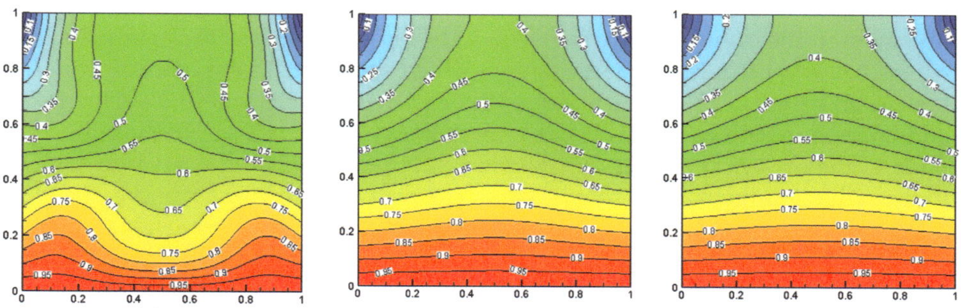

(a) (b) (c)

Figure 6.
Comparison of the steady state isotherms at Ra =10^5 and Pr = 0.025 for different Ha numbers in the case of linearly heated side walls. (a) Ha = 0 (b) Ha = 50 and (c) Ha = 100.

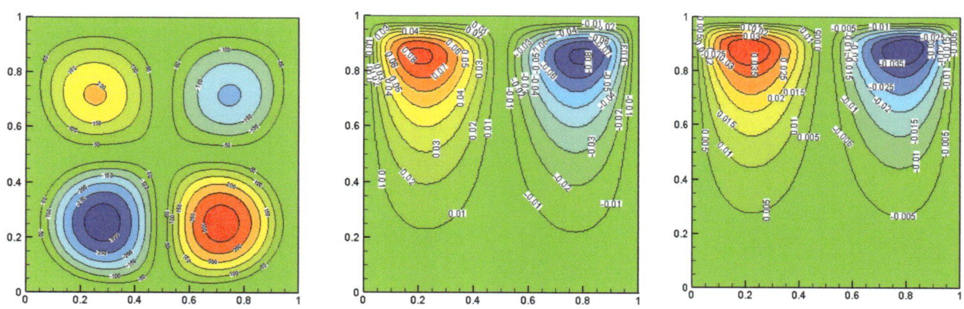

(a) (b) (c)

Figure 7.
Comparison of the steady state fluid velocity streamlines at Ra =10^5 and Pr = 0.025 for different Ha numbers in the case of linearly heated side walls.

Ra = 10^5. Numerical results in terms of flow and thermic structure show that the flow within the cavity takes place owing to the thermic buoyancy effects caused by the linearly heated walls, which is represented by the Ra number. The finding of Ra = 10^5 in the absence of magnetic field show that the flow is characterized by a multi-cellular behavior in which the re-circulating eddies or cells of relatively high velocity are formed within the enclosure. The main awareness is about the cells that occur at upper left and lower right side which are circulating in the anti-clockwise direction and the rest are circulating in the clockwise direction (see **Figure 7a**). Near the adiabatic wall of the cavity, stronger cells are formed. Provided that we implement the same boundary conditions for the left (west) and right (east) walls of the enclosure, the two pairs of cells close to each of the linearly heated walls are symmetric. As forecasted, because of the boundary layer effects, the temperature field is sketched by sharp drop in its value near the adiabatic wall of the enclosure while it increases away from it in order to reach its maximum at the bottom boundary. Besides, we highlight that the temperature contour maps are not horizontally uniform in the core region of the cavity. We infer that an in-plane magnetic field has the tendency to slow down the movement of the fluid in the confined enclosure. As an outcome, the recirculating cells stretch or elongated in the rising vertical direction. In addition, the same process remains as Ha number increases provided the lower cells merge with the primary recirculating eddies which are at the outset positioned near the upper wall for Ha = 50 and 100 (**Figures 6** and 7). Furthermore, vertical stretching of the cells occurs in such a way that the eyes of the cells get closer to the upper boundary. As a result, a significant reduction in the fluid movement in the cavity is spotted. Beyond, the braking effect of the magnetic field is discerned from the depicted circulation. When steady states are reached, temperature stratification becomes ruling as the strength of the magnetic field increases and temperature contour maps become more horizontally uniform in the core region of the cavity. The observed dynamic and thermic behavior converge on a quasi-conduction regime and this vanquish the overall heat transfer in the cavity. The observed dynamic behavior is very useful and effective in the metals semiconductor crystal growth industries [35] and in controlling melts (**Figures 6** and 7).

To the author's knowledge, studies have thus far addressed a mesoscopic approach in an MHD open cavity with linearly heated wall (**Figure 1**). The objective of the present chapter is therefore to predict dynamic and thermic heat transfer in a crucial engineering application. In this section, the cavity is investigated at the high Ra number of 10^5, high Ha number (Ha = 50) and Pr number of 0.025. The exit section of flow on the open boundary moves downward and the movement of the flow gets limited with the increment of Ha number that it influences heat transfer from the linearly heated wall to the cold open boundary (see **Figure 8**). The effect

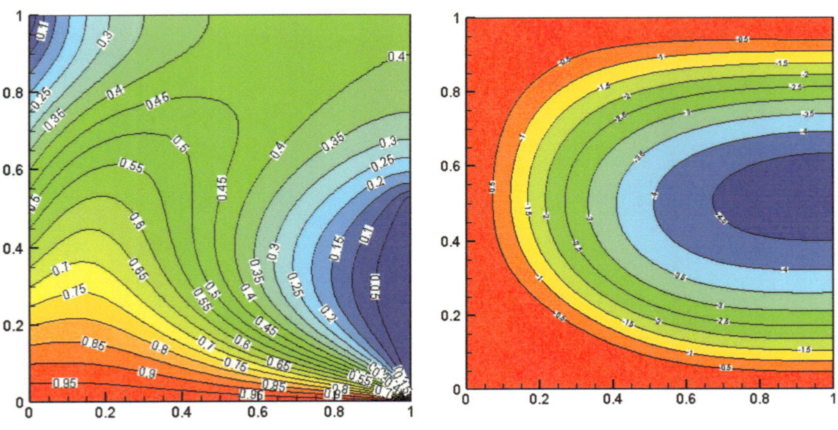

(a) (b)

Figure 8.
Steady state fluid velocity streamlines (a) and isotherms (b) at Ha = 50, Pr = 0.025 and Ra = 10^5 for linearly heated open cavity.

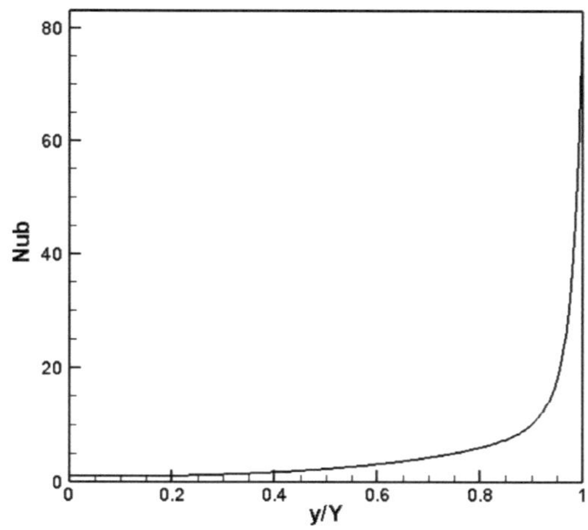

Figure 9.
Variation of local Nusselt number with distance at bottom wall in the case of linearly heated side wall for a MHD open cavity (Pr = 0.025, Ha = 50 and Ra = 10^5).

of the presence of the magnetic field is clear since the isotherms recede from the linearly heated left wall slowly and their gradient on the left wall declines extremely which it exposes the decrease in heat transfer in the open cavity. **Figure 9** illustrates the variation of local Nusselt number with distance at bottom wall in the case of linearly heated side wall for a MHD open cavity.

4. Conclusions

The present study considered steady state, laminar, two-dimensional MHD free convection within a liquid gallium filled square enclosure in the presence of in-plane magnetic field for different thermic boundary conditions. Lattice Boltzmann Method is utilized for simulation of this problem. The results of this method are validated in good agreement with previous numerical investigations. This investigation demonstrated ability of LBM for simulation of different boundary conditions at various elements affecting the stream. In addition this method helps to obtain stream function and isotherm counters smoothly. The governing equations are developed and solved by the LBM in the cases of linearly heated side walls and linearly heated left wall with cooled open right wall. In both cases, the upper wall of the enclosure is kept insulated. It was found that for the case of linearly heated side walls, a multi-cellular symmetric streamline behaviour was obtained for transverse applied magnetic fields. However, for the case of linearly heated left wall with cooled right wall, a two-cell (one primary and one secondary) feature was obtained. In general, the application of the magnetic field reduces the convective heat transfer rate in the cavity. Besides, the local Nusselt number at the bottom wall of the cavity exhibited oscillatory behavior along the horizontal distance for the case of linearly heated side walls whereas it increased continuously for the case of linearly heated left wall and cooled open right wall.

Nomenclature

C	lattice speed
c_i	discrete particle speeds
c_p	specific heat at constant pressure

F	external forces
f^{eq}	equilibrium density distribution functions
g^{eq}	equilibrium internal energy distribution functions
g	gravity
G	Buoyancy per unit mass
H	enclosure height
Ma	Mach number
Nu	Nusselt number
Pr	Prandtl number
Ra	Rayleigh number
T	temperature
x,y	cartesian coordinates
X	horizontal length of the cavity
Y	vertical length of the cavity
Ha	Hartmann number

Greek letters

ω_i	weighted factor indirection i
β	thermal expansion coefficient
τ	relaxation time
ν	kinematic viscosity
Δt	time increment
α	thermal diffusivity

Subscripts

av	average
H	hot
C	cold
b	bottom

Author details

Raoudha Chaabane[1,2]

1 Laboratory of Thermal and Energetic Systems Studies (LESTE) at the National School of Engineering of Monastir, University of Monastir, Tunisia

2 Preparatory Institute of Engineering Studies of Monastir (IPEIM), University of Monastir, Tunisia

*Address all correspondence to: raoudhach@gmail.com

IntechOpen

© 2019 The Author(s). Licensee IntechOpen. This chapter is distributed under the terms of the Creative Commons Attribution License (http://creativecommons.org/licenses/by/3.0), which permits unrestricted use, distribution, and reproduction in any medium, provided the original work is properly cited.

References

[1] Mohamad AA. Natural convection in open cavities and slots, numer. Heat Transfer Part A. 1995;**27**:705-716

[2] Polat O, Bilgen E. Laminar natural convection in inclined open shallow cavities. International Journal of Thermal Sciences. 2002;**41**:360-368

[3] Mohamad AA, El-Ganaoui M, Bennacer R. Lattice Boltzmann simulation of natural convection in an open ended cavity. International Journal of Thermal Sciences. 2009;**48**:1870-1875

[4] Mohamad AA, Bennacer R, El-Ganaoui M. Double dispersion natural convection in an open end cavity simulation via lattice Boltzmann method. International Journal of Thermal Sciences. 2010;**49**: 1944-1953

[5] Kennedy P, Zheng R. Flow Analysis of Injection Molds. Munich: Hanser; 2013

[6] Fidaros D, Grecos A, Vlachos N. Development of numerical tool for 3D MHD natural convection. Annex XX. pp. 73-74

[7] Sathiyamoorthy M, Chamkha A. Effect of magnetic field on natural convection flow in a liquid gallium filled square cavity for linearly heated side wall(s). International Journal of Thermal Sciences. 2010;**49**:1856-1865

[8] Alchaar S, Vasseur P, Bilgen E. Natural convection heat transfer in a rectangular enclosure with a transverse magnetic field. Journal of Heat Transfer. 1995;**117**:668-673

[9] Garandet J, Alboussiere T, Moreau R. Buoyancy drive convection in a rectangular enclosure with a transverse magnetic field. International Journal of Heat and Mass Transfer. 1992;**35**: 741-748

[10] Rudraiah N, Barron R, Venkatachalappa M, Subbaraya C. Effect of a magnetic field on free convection in a rectangular enclosure. International Journal of Engineering Science. 1995;**33**:1075-1084

[11] Cowley M. Natural convection in rectangular enclosures of arbitrary orientation with magnetic field vertical. Magnetohydrodynamics. 1996;**32**: 390-398

[12] Teamah M. Hydro-magnetic double-diffusive natural convection in a rectangular enclosure with imposing an inner heat source or sink. Alexandria Engineering Journal. 2006;**45**(4): 401-415

[13] Ozoe H, Okada K. The effect of the direction of the external magnetic field on the three-dimensional natural convection in a cubic enclosure. International Journal of Heat and Mass Transfer. 1989;**32**:1939-1953

[14] Ece M, Buyuk E. Natural convection flow under a magnetic field in an inclined rectangular enclosure heated and cooled on adjacent walls. Fluid Dynamics Research. 2006;**38**(5): 546-590

[15] Al-Najem N, Khanafer K, El-Refaee M. Numerical study of laminar natural convection in tilted enclosure with transverse magnetic field. International Journal of Numerical Methods for Heat and Fluid Flow. 1998;**8**:651-672

[16] Jalil J, Al-Taey K. MHD turbulent natural convection in a liquid metal filed square enclosure. Emirates Journal for Engineering Research. 2007;**12**(2):31-40

[17] Gelfgat A, Bar-Yoseph P. The effect of an external magnetic field on oscillatory instability of convective flows in a rectangular cavity. Physics of Fluids. 2001;**13**(8):2269-2278

[18] Aleksandrova S, Molokov S. Three-dimensional buoyant convection in a rectangular cavity with differentially heated walls in a strong magnetic field. Fluid Dynamics Research. 2004;**35**: 37-66

[19] Kahveci K, Oztuna S. MHD natural convection flow and heat transfer in a laterally heated partitioned enclosure. European Journal of Mechanics—B/ Fluids. 2009;**28**:744-752

[20] Kefayati GHR, Gorji M, Ganji DD, Sajjadi H. Investigation of Prandtl number effect on natural convection MHD in an open cavity by Lattice Boltzmann Method. Engineering Computations. 2013;**30**:97-116

[21] Bilgen E, Yedder RB. Natural convection in enclosure with heating and cooling by sinusoidal temperature profiles on one side. International Journal of Heat and Mass Transfer. 2007;**50**:139-150

[22] Sarris IE, Lekakis I, Vlachos NS. Natural convection in a 2D enclosure with sinusoidal upper wall temperature, numer. Heat Transfer Part A. 2002;**42**: 513-530

[23] Varol Y, Oztop HF, Pop I. Numerical analysis of natural convection for a porous rectangular enclosure with sinusoidally varying temperature profile on the bottom wall. International Communications in Heat and Mass Transfer. 2008;**35**: 56-64

[24] Saeid NH, Yaacob Y. Natural convection in a square cavity with spatial side wall temperature variation, numer. Heat Transfer Part A. 2006;**49**: 683-697

[25] Martinez D, Chen S, Matthaeus W. lattice Boltzmann magneto hydrodynamics. Physics of Plasmas. 1994;**6**:1850-1867

[26] Chaabane R, Askri F, Nasrallah SB. Parametric study of simultaneous transient conduction and radiation in a two-dimensional participating medium. Communications in Nonlinear Science and Numerical Simulation. 2011;**16**(10): 4006-4020

[27] Saha LK, Hossain MA, Gorla RSR. Effect of Hall current on the MHD laminar natural convection flow from a vertical permeable flat plate with uniform surface temperature. International Journal of Thermal Science. 2007;**46**:790-801

[28] Chaabane R, Askri F, Nasrallah SB. Analysis of two-dimensional transient conduction-radiation problems in an anisotropically scattering participating enclosure using the lattice Boltzmann method and the control volume finite element method. Journal of Somputer Physics Communications. 2011;**182**(7): 1402-1413

[29] Lamsaadi M, Naimi M, Hasnaoui M, Mamou M. Natural convection in a vertical rectangular cavity filled with a non-Newtonian power law fluid and subjected to a horizontal temperature gradient. Numerical Heat Transfer, Part A: Applications. 2006;**49**:969-990

[30] Chaabane R, Askri F, Nasrallah SB. Application of the lattice Boltzmann method to transient conduction and radiation heat transfer in cylindrical media. Journal of Quantitative Spectroscopy and Radiative Transfer. 2011;**112**(12):2013-2027

[31] Chaabane R, Askri F, Jemni A, Nasrallah SB. Numerical study of transient convection with volumetric radiation using an hybrid lattice Boltzmann BGK-control volume finite element method. Journal of Heat Transfer. 2017;**139**(9):092701

[32] Raoudha C, Askri F, Jemni A, Nasrallah SB. Analysis of Rayleigh-Bénard convection with thermal

volumetric radiation using lattice Boltzmann formulation. Journal of Thermal Science and Technology. 2017; **12**(2):JTST002. https://www.jstage.jst.go.jp/browse/jtst/12/0/_contents/-char/en0

[33] Mohamad AA. Applied Lattice Boltzmann Method for Transport Phenomena, Momentum, Heat and Mass Transfer. Calgary: Sure; 2007

[34] Succi S. The Lattice Boltzmann Equation for Fluid Dynamics and beyond, Clarendon Press. Oxford: London; 2001

[35] Series RW, Hurle DTJ. The use of magnetic fields in semiconductor crystal growth. Journal of Crystal Growth. 1991;**133**:305-328